Practical Liquid
Chromatography

S. G. Perry, R. Amos, and P. I. Brewer

Esso Petroleum Company, Limited
Esso Research Centre
Abingdon, Berkshire, England

Ⴔ *PLENUM PRESS · NEW YORK–LONDON · 1972*

Library of Congress Catalog Card Number 75-179760

ISBN 0-306-30548-8

© 1972 Plenum Press, New York
A Division of Plenum Publishing Corporation
227 West 17th Street, New York, N.Y. 10011

United Kingdom edition published by Plenum Press, London
A Division of Plenum Publishing Company, Ltd.
Davis House (4th Floor), 8 Scrubs Lane, Harlesden, London,
NW10 6SE, England

Printed in the United States of America

Practical Liquid Chromatography

Preface

This book is intended to provide a practical introduction to high-speed, high-efficiency liquid chromatography. It covers modern column technology (which has leapt into prominence only in the last five years) and relates this to the well-established thin-layer techniques.

The development of liquid chromatography has proceeded in fits and starts over many years and in alliance with various scientific disciplines. Liquid chromatography has for years fulfilled an effective role in various fields. Ion-exchange chromatography, for example, is particularly associated with the separation of the rare earths, and exclusion chromatography with the fractionation of naturally occurring materials like proteins and of synthetic polymers. Partition chromatography, especially in the form of paper chromatography, has been an indispensable tool in the study of biochemical systems, while its more recent adsorption counterpart, thin-layer chromatography, developed most rapidly within the pharmacognosic and pharmaceutical fields. Until recently, however, liquid chromatography has not played a prominent role in the field of industrial organic analysis.

Recent books have provided theoretical background to the mechanism and technique of liquid chromatography, but we felt there was a need for a practical book suitable for use at or near the bench of analysts in the organic chemical industry. To fulfill this need this book was written. We present no new theories but relate the theoretical conclusions of others to the practical needs for procuring effective separations of lipophilic substances in organic media. We consider, with a practical emphasis, the mechanics of adsorption, partition, and exclusion chromatography and describe the techniques of thin-layer and high-performance column chromatography based largely on our own experiences. We hope this book will help organic chemists to solve their separation problems, and encourage the further development of high-

v

efficiency liquid chromatography to share the same uniquely successful role presently enjoyed by gas chromatography.

This is not intended to be a comprehensive reference book. In fact, references have been cut to a minimum. Together, however, with a select range of further reading the reader is directed to most of the significant publications which have so far appeared.

While every endeavor has been made to make lists of manufacturers comprehensive, it is inevitable, especially in the presently rapidly expanding field of high-performance column chromatography, that they will rapidly become obsolete. We apologize for any omissions.

We would like to thank Professor A. I. M. Keulemans, who has done us the honor of writing the Foreword to this book. He it was who wrote the first widely read book on gas chromatography, which undoubtedly did much to stimulate interest and work in the field. We will feel our efforts have been worthwhile if this text provides an even remotely comparable stimulus for liquid chromatography. We are also greatly indebted to various authors and publishers for their permission to reproduce certain tables and figures, which are identified at the appropriate stages throughout the book.

Finally, we would like to thank the management of the Esso Research Centre, Abingdon, for their active encouragement to write this book and for providing facilities for preparation of the manuscript.

S.G. PERRY
R. AMOS
P.I. BREWER

Abingdon, September, 1971

Foreword

A Dutch proverb says "Goede wijn behoeft geen krans" and the nearest English translation may be: "Good wine needs no recommendation." It gives me great pleasure to write a foreword to a book written by Dr. Perry and his colleagues. On the one hand he belongs to the youngest "old hands" from the time of the explosive development in gas chromatography; on the other hand he has played a pioneering role in the recent renaissance of liquid chromatography. Mr. Amos has played a big part in popularizing use of thin-layer techniques in the analysis of petroleum products, and Mr. Brewer is the author of one of the first papers on the use of exclusion chromatography for separation of synthetic polymers. The authors do not shy from theory but, being employed in an industrial laboratory, they are first of all practical men. Not only are the authors well known but through their many contacts personally know the older and the present generation of publishing chromatographers. I have had the privilege to have Dr. Perry working for one year as a post doctoral research chemist at our University Laboratories. I was, therefore, in the position to learn to appreciate his many outstanding abilities. Not the least of his abilities are his balanced judgment and his eloquence. The present book clearly shows these abilities. It deals with theory only where this is necessary. The great potentialities but also the limitations of the chromatographic techniques labeled under the collective name "Liquid Chromatography" are treated in a most objective way. The book gives, what is badly needed, a practical guide for present and prospective workers in the fascinating field of liquid chromatography. Of course a feature of the book is its bibliography. The present possibilities offered by computers greatly facilitate the preparation of complete bibliographies, but here the authors have made a critical selection. The book does not need a foreword to find its way to the desks of the practicing chromatographers where it soon will show itself an indispensable guide. To me this foreword means an open confirmation of cordial friendship between past and future, between two generations of enthusiastic chromatographers.

A.I.M. KEULEMANS
University of Technology
Eindhoven, Holland

Contents

Chapter 1

Liquid Chromatography—The Background

1.1. PERSPECTIVE

Liquid chromatography has an impressive history, stretching back into the 19th century, which has been well documented elsewhere[1]. As an indispensable technique for the separation and analysis of organic substances it has made enormous strides in the past decade or so. Prior to 1958 column chromatography was conducted with comparatively large samples separated in, often short, almost invariably wide-bore, columns through which the developing liquid percolated under gravity. Thin-layer techniques were unknown outside a small group of workers[2], although paper chromatography was widely practiced.

Since 1958, thin-layer chromatography has emerged as the simplest, lowest cost, and comparatively most efficient method available for the separation of the less volatile components of organic mixtures. Many laboratories make extensive use of it for research and quality control in the whole field of industrial (and academic) organic chemicals. More recently the techniques of column chromatography have been radically rethought, to the point where they are now directly comparable in terms of ease, efficiency, and speed with those of gas chromatography.

With this book we seek to provide an adequate and up-to-date account of the theory and practice of liquid chromatography as applied to the analysis of mixtures of organic substances. The emphasis throughout is on the practical aspects of the subject but with the basic elements of theory included so that the reader can appreciate the reasons underlying recommended practice.

1

1.2. PLAN OF THE BOOK

Liquid chromatography is an enormous subject which, furthermore, is passing through a phase of rapid progress. Therefore we have sought to concentrate on those aspects of particular interest to practicing organic analytical chemists in industry. We have also attempted to provide an understanding of the framework of the techniques so that this book will not rapidly become out-of-date, notwithstanding the rapid advancement of liquid chromatography at the time of writing.

In the remainder of this chapter we summarize the variants of liquid chromatography and define or explain some of the more important terms encountered in the literature and the laboratory.

Chapter 2 seeks to provide an adequate explanation of the dynamic processes which occur in a bed of stationary phase through which a liquid is flowing. These processes ultimately limit the degree of separation that can be achieved. They cause "band spreading" in all chromatographic processes. The theories which have been evolved to relate operating parameters to band spreading are outlined and their practical consequences emphasized.

Each of three succeeding chapters then considers the essentials of major subdivisions of liquid chromatography, namely adsorption, partition, and exclusion. In each chapter an explanation is given of the mechanism of selectivity and the separation achievable with the particular system concerned. The materials, especially the stationary phases, are discussed in depth—as always with the important practical points emphasized.

The next two chapters deal in turn with the apparatus and techniques of, first, thin-layer chromatography and, then, column chromatography.

A final brief chapter seeks to put into perspective the current status of liquid chromatography.

The book is intended as a practical guide, and for this reason we have limited the references cited to a relatively small number and we have backed these up with lists of articles suggested for further reading. In general we have sought to avoid quoting historical data and the derivation of equations from theory, as these topics are already very adequately treated in the literature. Also we have almost completely avoided referring to specific separations; an abstract service,* journals,† and books[1,3] provide information of this type. In short this book seeks to explain the "what, why, and how" of the techniques and materials used in modern liquid chromatography;

---

*The Journal of Chromatography regularly publishes bibliographies covering column and thin-layer chromatography.

†Journals regularly publishing papers on liquid chromatography and its application include: Analytical Chemistry; The Analyst; The Journal of Chromatographic Science; The Journal of Chromatography.

paper chromatography and ion-exchange chromatography (in partly or wholly aqueous media) are not considered in this book.

1.3. CHROMATOGRAPHY AND ITS BASIC VARIANTS

Chromatography is a process whereby different types of molecules are separated one from another. A sample mixture is introduced onto a bed of *stationary phase* and swept through it by a fluid at a rate dependent on the mutual interactions of sample components with the stationary phase and the fluid (*mobile phase*). Generally these mutual interactions differ in magnitude for the different sample components so that their rate of passage through the stationary phase bed differs and separation is thus achieved.

When the fluid mobile phase is a liquid the process is termed *liquid chromatography* and it is this subject with which the present book is concerned. Other types of fluid which are used are gases (gas chromatography) and substances above their critical points (supercritical chromatography).

The mobile phase introduced at the front end of the stationary phase bed is often referred to as the *eluent* and the eluent plus solutes leaving the end of the bed is sometimes termed the *eluate*.

Stationary phases promote separation of molecules if they possess one, or more, of four basic functional characteristics: (i) the power physically to sorb solutes from solution; (ii) the power chemically to sorb solutes from solution; (iii) the ability to dissolve solutes when contacted with solutions in an immiscible solvent; (iv) a porous structure which can retain some, and reject other, solutes on the basis of solute size or shape.

Each of these characteristics is the basis for a widely recognized variant of liquid chromatography and each will now be described in outline.

1.3.1. Adsorption Chromatography

Adsorption chromatography (also referred to as liquid–solid chromatography, LSAC, and LSC) is based primarily on differences in the relative affinity of compounds for the solid adsorbent used as stationary phase. Affinity is determined almost entirely by polar interactions. This means that polar groupings in the molecules to be separated exert a much greater effect than nonpolar hydrocarbon chains and adsorption chromatography therefore tends to separate mixtures into classes characterized by the number and type of polar groups.

1.3.2. Ion-Exchange Chromatography

Ion-exchange chromatography is based on the differing affinity of ions in solution for sites of opposite polarity in the stationary phase. It is a process largely confined to media of high dielectric constant, in which ionic species are stable. Most ion-exchange separations are carried out in essentially aqueous media and the principal fields of application are in inorganic chemistry. Some work has been carried out in mixed aqueous organic solvents[4], but in this book discussion of the application of ion-exchanging stationary phases is confined to situations in which they are effectively acting as adsorbents possessing specific functional affinities.

1.3.3. Partition Chromatography

Partition chromatography (liquid–liquid chromatography, LLC) is based on the relative solubility (or, better, distribution) characteristics of solutes between the mobile phase and a liquid (solvent) phase held stationary by impregnation on a porous, inert *support*. Most commonly the stationary phase is more polar than the mobile phase; in some circumstances, however, it is advantageous to reverse the roles so that the stationary phase is less polar. This variation is known as *reversed-phase partition chromatography*.

In principle LLC offers a wide range of selectivity effects as the relative nature of the two liquid phases is varied. In practice, as we will see, the choice of liquid phases which can be used is restricted. Actual separations essentially depend on the balance between polar and apolar groups in the solutes and the two liquid phases. Thus separations by compound type or chain length can be obtained.

1.3.4. Exclusion Chromatography

Exclusion chromatography, which is also known as gel permeation (GPC), gel filtration, or molecular sieving, is based on the ability of controlled-porosity materials to sort and separate sample mixtures according to the size and shape of the components of the sample. It has developed along two parallel lines with surprisingly little cross-fertilization. One line has been the application of hydrophilic polymers such as cross-linked dextrans to the separation of biological materials in primarily aqueous media. The other has been the use of synthetic materials, polymers, or inorganic porous substances for the separation of industrial organic chemicals, especially plastics and polymers. These latter separations are almost invariably performed in lipophilic solvents and in this book attention will be concentrated on them.

Actual separations may involve a combination of the above basic mechanisms, for example, in liquid–solid chromatography adsorption can take place physically, e.g., by dipole–dipole interaction, or chemically, e.g., by ion exchange or complex formation. In addition, the solid can sorb a portion of the mobile phase, especially if this contains water or some other polar solvent, to form an organized stationary liquid into which solutes can partition.

1.4. DEVELOPMENT METHODS IN CHROMATOGRAPHY

The process by which solutes are carried through the stationary phase by the mobile phase is called *development*. One should be careful to avoid its confusion with the act of spraying thin-layer chromatograms with reagents which form colored derivatives and so reveal the presence of separated components. This latter process is often referred to as "development" (of colored spots); it is better to refer to it as *visualization* or *revealing*.

Chromatograms can be developed by three different methods, *elution, frontal analysis, and displacement*. Of these methods elution is by far the most widely used. In fact frontal analysis is never used as a practical analytical method. Displacement techniques are occasionally used; in particular a very important and widely used method of hydrocarbon type analysis, standardized by the American Society for Testing Materials and the Institute of Petroleum, is based on displacement development. The following sections describe the three methods briefly.

1.4.1. Elution

Consider a mixture of solutes A and B, placed initially at one end of a bed of stationary phase, and suppose that B is more strongly retained than A. If a mobile phase, less strongly retained than either A or B, is caused to pass through the bed, then it will wash A and B through at different speeds according to the degree of retention of the two solutes. If the difference in migration rates is sufficiently great, then A and B, which are initially superimposed, will gradually separate to form two distinct zones with pure eluent separating them, A moving ahead of B. If the sorption isotherms are linear (see Chapter 3) then a plot of concentration of the solutes along the length of the bed will appear as in Fig. 1.1, each solute having a Gaussian concentration profile.

1.4.2. Frontal Analysis

In this case, the mixture is fed continuously into one end of the stationary phase bed and caused to flow toward the other end. Again B is more strongly retained than A so that the solute front will become depleted in B and eventually pure A will emerge at the other end. Meantime the bed will become saturated with B and it too will then be washed forward along with A so that the mixture will flow through the bed with its original composition unchanged. Thus, by this method of separation, it is only possible to recover a pure sample of component A.

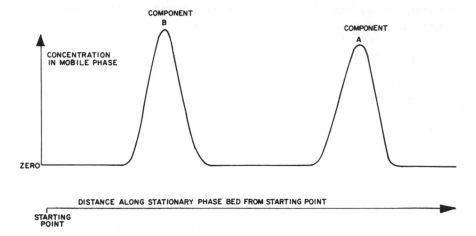

Fig. 1.1. Elution of binary mixture.

1.4.3. Displacement

As in elution, a small sample is placed at one end of a bed of stationary phase, but this time the mobile phase which is fed in is more strongly retained than either A or B. It therefore displaces these components through the bed, and since B is more strongly retained than A, the latter moves ahead of B. Intermediate zones in which the composition changes from pure A to pure B and from pure B to pure mobile phase also exist.

Thus, in contrast to frontal analysis, a portion of both components may be obtained in a pure state.

1.5. BASIC EXPERIMENTAL ARRANGEMENTS

The bed of stationary phase is either contained in a column or else spread out as a thin, open bed. In the latter case it is either supported on a glass, metal, or plastic backing as in thin-layer chromatography, or else the bed is self-supporting as in paper chromatography. The basic *modi operandi* of the column and layer techniques are as follows:

1.5.1. Column Chromatography

A typical experimental setup is shown schematically in Fig. 1.2. The stationary phase is packed into a tube of glass, plastic, or metal. The sample under examination is placed at the top of the packed tube and the mobile phase is passed through the column from an eluent reservoir. The flow of mobile phase is provided either by gravity flow or, if higher pressures are needed, by a pump. The sample components emerge from the end of the column as a dilute solution in the mobile phase and may be monitored by a suitable detector or collected.

1.5.2. Layer Chromatography

The stationary phase is spread out as a thin layer on a supporting surface such as glass, plastic, or metal. The sample is applied near one edge of the layer, which is then dipped into the mobile phase. Solvent flows through the stationary phase by capillary action (ascending or horizontal chromatography) or by gravity flow (descending chromatography). Sample components migrate through the bed, but chromatography is generally stopped before the solutes reach the outer edge of the layer. The separated zones are either examined *in situ* or else are removed for further investigation.

1.6. MEASURES OF THE EFFECTIVENESS OF CHROMATOGRAPHIC SEPARATIONS

Some quantitative measure of the degree of separation achieved in a

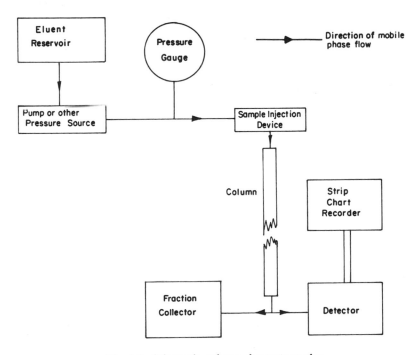

Fig. 1.2. Schematic column chromatography.

chromatographic process is desirable for at least two reasons. First, it would enable the effect of varying the different operating parameters to be assessed objectively and thus guide the experimenter in adjusting these parameters to achieve even better separations. Second, knowledge of the effectiveness of a column or thin layer to separate a given sample mixture would enable a prediction to be made of its ability to separate a different sample.

Two quantities are widely used as such a measure, the "height equivalent of a theoretical plate" (HETP) and the resolution (R). These terms can be related through such properties as distribution coefficients. However, in this book we are not to be concerned with these relationships and the reader is referred to the several admirable texts which deal thoroughly with such matters (see the list of further reading at the end of this chapter). We need, however, to appreciate the significance of the two above-mentioned terms and how to calculate their magnitude.

1.6.1. HETP

This term, often further abbreviated to H, is basic to the derivation of quantitative theories of chromatographic mechanisms and will be extensively used in the next chapter. It derives from theories of distillation, a theoretical plate being a column zone in which vapor and liquid are in equilibrium. The length, or height, of this zone is the HETP. The development of chromatographic theory by Martin and Synge followed lines similar to the so-called plate theory, and all more recent treatments have also used H as the funda-

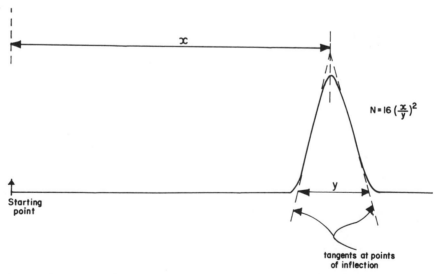

$$N = 16 \left(\frac{x}{y}\right)^2$$

Fig. 1.3. Calculation of number of theoretical plates from an elution chromatogram.

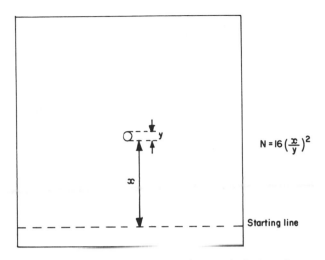

Fig. 1.4. Calculation of number of theoretical plates from thin-layer chromatogram.

mental measure of column or thin-layer performance. Figure 1.3 is an elution chromatogram as recorded by a detector at the exit of a column of length L. The *number* N of theoretical plates generated by the column is given by

$$N = 16 \, (x/y)^2$$

and therefore the length of column per plate (which is the HETP) is given by $H = L/N$. It should be noted that there are several other methods for calculating N which may be encountered: the one given here is probably the most frequently used and easiest to measure.

It will be noted that the expression for N is essentially a function of the quotient of solute retention and solute peak width. The thin-layer analog is shown in Fig. 1.4, where

$$N = 16 \, (x/y)^2$$

In this case, however, $L = x$ and therefore $H = y^2/16x$.

1.6.2. Resolution

The HETP is a measure of overall performance but its value varies with the retention time of a solute. A practical disadvantage it suffers is that in using it to compare different columns their ability actually to separate substances does not increase in parallel with an increase in HETP. Therefore a direct measure of the ability to *separate* two components is very useful for comparative purposes, and *resolution* is a most convenient parameter to use.

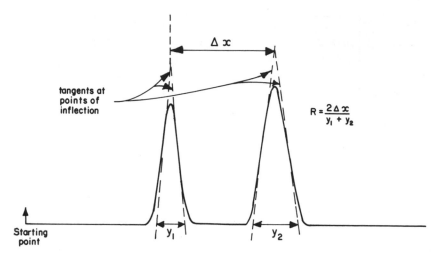

Fig. 1.5. Calculation of resolution in elution chromatography.

Resolution is becoming used increasingly as the criterion by which gas-chromatographic columns are accepted in standardized methods; doubtless it will play a similar role for liquid chromatography in the future. It should be noted that resolution, unlike HETP, includes a selectivity contribution, the separation of two peak maxima, as well as a performance factor, the peak width.

For the two separate components in the elution chromatogram shown in Fig. 1.5 the resolution R is given by

$$R = 2\,(\Delta x)\,/\,(y_1 + y_2)$$

As thus defined R can be regarded pictorially as the maximum number of additional resolved components which could be accommodated between the two components indicated.

1.7. RETENTION PARAMETERS

In a given chromatographic system the time taken (and the volume of mobile phase required) to elute a specific substance from the stationary phase bed is constant. It follows therefore that the time (or volume) needed to elute one substance relative to another particular compound is also constant. These facts lead us to consider two aspects of retention in chromatography: (i) the various methods of describing retention and their interrelation and (ii) the use of retention as a means of identification of solutes.

1.7.1. Measures of Retention

The retention volume V_r of a solute is the volume of mobile phase needed to elute it through the bed of stationary phase. The "dead volume" V_0 of a bed of stationary phase is the volume of mobile phase which it contains when saturated. The *partition coefficient K* of a solute is the ratio of its concentration in the stationary phase to that in the mobile phase at equilibrium; this ratio is also known as the *distribution coefficient.* These quantities are related by the expression

$$V_r = V_0 + KV_s \qquad (1.1)$$

where V_s is the volume of stationary phase in the bed.

For a given system it is obvious that the retention of a substance may be expressed in *absolute* terms (as the volume of eluent to elute it), in *specific* terms (as the volume of eluent per gram of stationary phase to elute it), or in *relative* terms (as the ratio of, say, absolute retention volume of solutes to that of a chosen reference solute). Both absolute retention volume and relative retentions are widely used to characterize the behavior of solutes in liquid chromatography.

Another retention parameter used, particularly in flat-bed (i.e., paper or thin-layer) chromatography, is the R_F *value.* The R_F value of a solute is the ratio of the distance it has moved in a given time to the movement of the front of the mobile phase. Figure 1.6 shows how the R_F value of a solute in TLC (thin-layer chromatography) is measured.

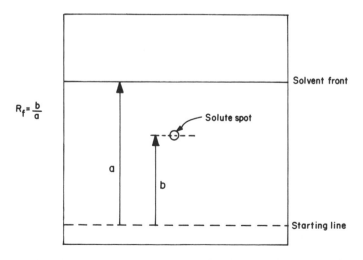

Fig. 1.6. Calculation of R_F value in thin-layer chromatography.

1.7.2. Retention and Identification

Expression (1.1) explicitly shows that in a fixed system V_r (and therefore R_F) is a constant for a particular solute; V_r can therefore be likened to other physical constants of the solute and be used to identify that solute. Thus, although chromatography is primarily a separation technique, it is possible to identify the separated components of a complex sample by their retention characteristics. Such identifications can be confirmed by isolating the eluted solute and, for example, obtaining its mass or infrared spectrum. Because several solutes may have the same V_r (or R_F) in a given system, or because these quantities have not been determined with known substances, it is not uncommon to use chemical and spectral data obtained on solutes recovered from a separation to identify them.

The retention of solutes is related to their chemical structure, as of course is their partition or distribution coefficient. For example, in liquid partition chromatography the retention of the ith member of a homologous series is given by

$$\log V_{r_i} = mN_i \tag{1.2}$$

where m is a constant and N_i the number of carbon atoms in the homolog. One can therefore predict retentions from the known behavior of other members of a homologous series.

It is also sometimes possible to make use of the fact that introduction of a particular functional group into a molecule leads to a definite change in V_r, namely ΔV_r. From tabulations of ΔV_r values observed for various functional groups in one or two different chromatographic systems conclusions can be drawn about the chemical nature of "unknown substances."

The point made in the preceding paragraphs and needing to be emphasized is that retention is a characteristic property of solute and varies in a systematic way with variation in structure.

REFERENCES

1. Heftmann, E. (ed.), *Chromatography*, Reinhold, New York (1967).
2. Pelick, N., Bolliger, H.R., and Mangold, H.K., The history of thin-layer chromatography, *Advan. Chromatog.* **3**, 85 (1966).
3. Stahl, E. (ed.), *Thin Layer Chromatography*, Springer-Verlag, Berlin (1969).
4. Webster, P.V., Wilson, J.N., and Franks, M.C., *J. Inst. Pet.* **56**, 50 (1970).

FURTHER READING

Authoritative reviews on various aspects of liquid chromatography are to be found in two regular publications:
Chromatographic Reviews, Lederer, M. (ed.), Elsevier, Amsterdam.
Advances in Chromatography, Giddings, J.C., and Keller, R.A. (eds.), Marcel Dekker, New York.

Chapter 2

Band Spreading in Chromatography

2.1. INTRODUCTION

Many outsiders with a scientific training looking in at the practice of chromatography must regard the exercise as an art or mystique, rather than as a science. There is no doubt that empiricism and experience frequently provide the dominating guidelines in setting up an experiment; theoretical considerations, even of a rudimentary nature, only infrequently play a major part in the design and operation of a chromatographic separation. This situation is even more pronounced in liquid chromatography than in the newer, but enormously more successfully applied technique of gas chromatography. Indeed, despite its seventy or eighty years of history, liquid chromatography has made very little progress, the materials and equipment in use today (at least for most adsorption work in columns) differing rather little from that of Day and Tswett.

The intention in writing this book is to enable the reader to solve practical separation problems as successfully as present knowledge will allow. The availability of sophisticated instrumentation and of recently developed materials (such as porous polymers, controlled-surface-porosity supports, and ion-exchange resins) with well-defined chemical and physical properties makes possible separations which hitherto were impossible; these matters will be discussed elsewhere in this book. In this chapter we will look at the underlying processes which occur in the elution of solutes by liquids through beds of solids. Many of the experimental facts on which present knowledge of this area is based come from the numerous studies made with gaseous mobile phases in gas chromatography, while chemical engineering studies on adsorbent and catalyst beds and geochemical investigations of oil-field structure have also made significant contributions. In the past two or three years some basic work on liquid chromatography has been carried out. It has generally confirmed that the conclusions drawn from work in the fields referred to above are valid in liquid chromatography, too.

The object of carrying out a chromatographic experiment is generally to achieve a separation, perhaps of two particular substances, perhaps of all components in a complex mixture. The objective may be analytical; that is, the separation has to be sufficient that the different components are separately recognizable and measurable. On the other hand, a preparative separation may be required, to provide a more-or-less pure fraction or substance for further work. In this case the requirement is for virtually complete separation of rather larger quantities than are necessary for successful analytical chromatography.

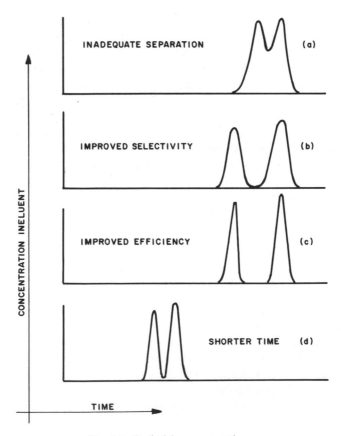

Fig. 2.1. Optimizing a separation.

There are two routes to the achievement of chromatographic separation. The first, which will be discussed in detail in subsequent chapters, can be termed the selectivity route. This way the two immiscible phases are chosen so that the components required to be separated have sufficiently different retardation that they may be relatively easily separated. The other approach can be called the performance route; the making of the stationary phase bed and the operation of forcing the liquid phase through it are very carefully conducted in order that materials with even slightly different distribution coefficients are, nevertheless, adequately separated. In practice a combination of the two is usually employed. The two routes are illustrated diagrammatically in Fig. 2.1. In Fig. 2.1(a) the chromatographic system fails to achieve an adequate separation. By changing, for example, to a more selective stationary phase a good separation is obtained in the same time [Fig. 2.1(b)]. Using the same phases, but with careful design and operation, much sharper peaks and, consequently, better separation in the same time are achieved [Fig. 2.1(c)]. This greater separation, achieved through careful attention to design and operation, can be traded so that a separation as good as that of Fig. 2.1(b) can be obtained in much shorter time [Fig. 2.1(d)].

In this chapter we will discuss the factors which control the performance or resolving power of a liquid chromatographic separation.

2.2. FACTORS LEADING TO BAND SPREADING

A quantity of a solute introduced into the upstream end of a packed chromatographic column will occupy a finite length of the column. During its passage through the column it will become more diffuse (Fig. 2.2). This band-spreading effect always occurs and, of course, militates against high resolving power. The greater the extent of band spreading of two adjacent components in a chromatographic system, the more difficult it is to separate them. Therefore it is important to recognize those aspects of packing design and geometry and of operations which influence band spreading. Further, we wish to put the effects onto a quantitative basis and draw from the expressions derived the conclusions which will enable us to design and conduct chromatographic separations of adequate resolving power for the task in hand.

The understanding of the chromatographic process has advanced greatly in the last three decades. This progress is examined and the evolution of the concepts of zone spreading discussed by Giddings, who is responsible for presenting the most fruitful exposition of chromatographic processes to date[1]. Among the milestones along the way perhaps the first was a paper by Wilson[2], which drew attention to the effect of mobile phase velocity. He noted that the leading edge of a band of solute migrated through an adsorbent bed at a faster rate than the main bulk of solute because of a low

rate of adsorption and that the trailing edge moved more slowly on account
of a low rate of desorption. Wilson pointed out that the adverse effects of
these low rates of mass transfer between fixed and moving phase could be
diminished by reducing the flow rate of the latter, but he also noted that at
lower flow rates the undesirable effects of diffusion become increasingly
pronounced. Thus before 1940 we see that three important features of chromatography had been recognized: (i) the influence of rate of mass transfer
between phases on band broadening; (ii) the effect of diffusion; (iii) the
realization that the influence of flow rate on mass transfer and diffusion
effects were opposed, so that the optimum mobile phase velocity would have
to represent a compromise.

Wilson's work was followed quickly by Martin and Synge's classic
paper[3] of 1941, in which the "plate theory" of chromatography was introduced. The essence of this theory is that a chromatographic column is
imagined to consist of a number of layers in each of which equilibrium
distribution of the solute exists between the fixed and moving phase. Such
layers are termed "theoretical plates" and their thickness H is the "height
equivalent of a theoretical plate" (HETP). Assuming (and it must be noted
that these assumptions can be seriously in error) that the phase equilibria are
independent of concentration and that diffusion effects are absent, a mathe-
matical treatment of the progress of a band of solute through a column leads
to the picture of the shape and width of the eluted band in terms of the normal
error curve (Fig. 2.3).

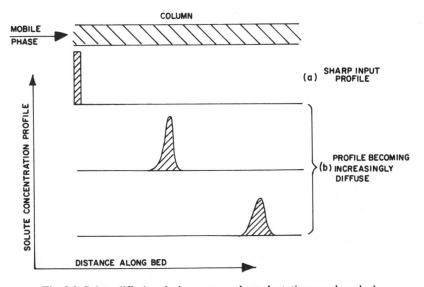

Fig. 2.2 Solute diffusion during passage through stationary phase bed.

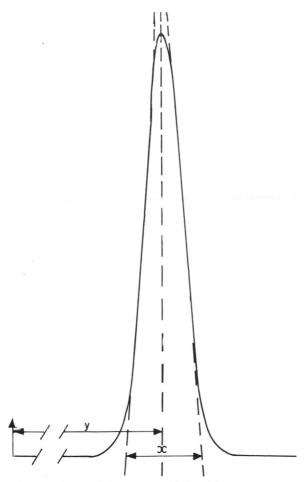

Fig. 2.3. Shape of chromatographic band from plate theory.

The HETP of a column can be calculated from the expression

$$H = Lx^2 \,/\, 16y^2$$

where L is the column length and x and y have the significance shown in Fig. 2.3.

The plate theory is a convenient tool for assessing the performance of a column, but in itself affords no explanation of real behavior in the column. Martin and Synge recognized the need for an intermediate mobile phase velocity, as Wilson had done before them, and also concluded that HETP was directly related to the square of the diameter of the column packing. Other factors recognized by Martin and Synge to be important were the serious effect of solute diffusion in broadening bands, the inconstancy of partition coefficients at higher concentrations, and the loss of efficiency through nonuniform flow through the column.

Thus by the early 1940's many of the crucial factors on which good chromatographic performance depend had been identified, at least in qualitative terms. In the next two decades detailed progress occurred, culminating in the random-walk and generalized nonequilibrium theories of Giddings. Before turning to those theories, which we will describe and discuss at some length, one other important milestone needs to be examined, the expression familiar to gas chromatographers as the "van Deemter equation." This equation was developed by van Deemter *et al.*(4) to relate HETP to flow velocity, taking mass transfer, longitudinal diffusion, and nonuniform flow paths into account.

The "van Deemter equation" may be written

$$H = 2\lambda d_p + \frac{2\gamma D_m}{v} + \frac{8}{\Pi^2}\left[\frac{k'}{(1 + k')^2}\right]\frac{d_f^2 v}{D_s} \qquad (2.1)$$

where λ is a dimensionless parameter related to packing irregularities; d_p is the mean particle diameter; γ is another parameter, like λ, related to the varying flow; D_m is the molecular diffusion coefficient of the solute in the mobile phase; v is the mean mobile phase velocity; k' is the partition ratio (i.e., amount of solute in the stationary phase divided by the amount in the

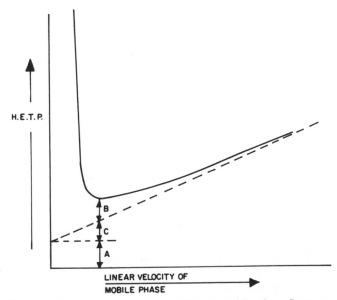

Fig. 2.4. Relation between plate height and mobile phase flow rate.

mobile phase at any instant); d_f is the stationary liquid film thickness; and D_s is the molecular diffusion coefficient of the solute in the stationary phase. Equation (2.1) is often simplified

$$H = A + (B/v) + Cv \qquad (2.2)$$

and has the form of a hyperbola, as shown in Fig. 2.4. There is no doubt that this relationship has guided most chromatographers in their efforts to make high-efficiency gas chromatographic columns. It has materially contributed to the enormously successful exploitation of GC (gas chromatography) and can still be used effectively for that purpose. Also, with suitable adjustment of numerical values of, for example, diffusion coefficients, its use can guide the construction and operation of high-efficiency liquid chromatographic columns.

We see at this point that there are a number of theories of chromatography all of which, directly or indirectly, tell us something of the process. We have briefly touched upon one or two of the more important of these theories and, with their help, identified some important chromatographic parameters. The object of this chapter, in a book directed to the practical man with analytical problems to solve, is to provide a clear picture of the mechanism of chromatography such that it can be applied at the bench to the solution of those problems. Therefore we are to utilize Giddings' very informative and easily understandable approach to explain the mechanisms of zone spreading in chromatography. These considerations occupy the remainder of this chapter. For a full treatment the reader is referred to Giddings' excellent book[1].

2.3. THE "RANDOM-WALK" CONCEPT

The following discussion is based on a model system in which the stationary phase is packed into a tube with a high aspect ratio (i.e., length much greater than diameter). The mobile liquid phase percolates through the irregularly packed bed of, generally, irregular-shaped particles (Fig. 2.5). The rate at which a molecule passes through the column is governed principally by its affinity for the stationary phase: the greater this is, the longer the molecule is immobilized. Thus the design of selective liquid chromatographic separations involves the choice of phase pairs to maximize differences in affinity toward solute molecules. This aspect will be discussed later in the book.

Superimposed on this average rate of travel of a molecule of a particular type through the column, there are the various small deviations from the mean value brought about by a finite rate of mass transfer of the solute between the phases, the diffusion of the solute in both the mobile and the stationary

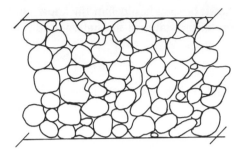

Fig. 2.5. Magnified cross section of packed bed

phases, and the range of different flow path lengths through the stationary phase which exist because of the irregularity of the packing of the bed.

These are the processes we have already identified as being concerned with band broadening and are the features we are to examine with the aid of the random-walk concept.

At the molecular level, movement is random—a cloud of smoke disperses in all directions in still air, while in a light breeze the dispersion is superimposed on a drift with the wind. Similarly, a sharp "plug" of solute in a chromatographic column will tend to disperse under the influence of intermolecular collisions so that it will take up a Gaussian distribution of concentration centered upon its mean position.

In considering the consequences of the random nature of mass transfer one should distinguish three rates of movement of solute molecules through the column. There is the rate at which they move when in the mobile phase, V_m, and the rate when in the stationary phase, which is zero. In addition there is the average rate of movement while in the column, V_{av}:

$$V_{av} = R_r V_m \tag{2.3}$$

where R_r is the retention ratio, the proportion of solute molecules in the mobile phase at any instant. The important point to note therefore is that as soon as a solute molecule passes into the stationary phase it falls behind the band center, while when it desorbs into the mobile phase it moves more rapidly than the center. The progress of a molecule through the column is seen to be a succession of random stops and starts, sometimes ahead of, sometimes trailing, the band center. The net effect is a symmetric dispersion of the molecules of a given solute about the mean position, the extent of the dispersion increasing with the number of "stops" and "starts" and also with the mobile phase velocity.

In a fixed bed of irregularly shaped and packed particles flow paths for the mobile phase are more or less irregular; some may be open while others are constricted. Accordingly there will be very considerable local variations in mobile phase velocity, while the actual paths followed by different solute molecules in the mobile phase will vary in length. This effect, which also leads to a dispersion of solute molecules about their mean position, is often referred to, particularly in gas chromatography, as "eddy diffusion."

Such are the random processes occurring in liquid chromatography and we must now apply some simple statistics to them to translate the qualitative picture into measurable and controllable parameters which will enable us to perform chromatographic separations as effectively as possible.

2.4. SIMPLE MATHEMATICS OF THE RANDOM WALK

Imagine an assembly of solute molecules lined up in a fluid. Erratic molecular movements will lead to steps of widely differing distances on either side of the starting line. The concentration profile after n steps of average length l have been taken is a Gaussian curve (Fig. 2.6) for which the standard deviation σ is given by

$$\sigma = ln^{1/2} \tag{2.4}$$

(the square relationship between σ and n reflects the fact that successive random walks by a molecule may be in the same or in opposite directions, the latter circumstance diminishing the net dispersion).

In chromatography different random processes are occurring simultaneously; the overall relationship between the net standard deviation σ and

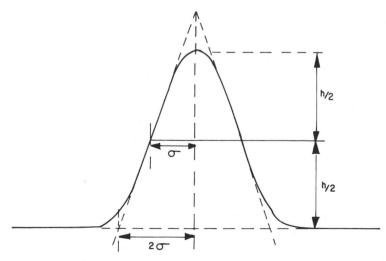

Fig. 2.6. Elution peak profile.

those of the separate processes, σ_1, σ_2, ... is

$$\sigma^2 = \sum_1^i \sigma_i^2 \tag{2.5}$$

A further important relationship derived from the plate theory is

$$H = \sigma^2/L \tag{2.6}$$

where H is the HETP and L is the column length.

In considering molecular diffusion Einstein's equation

$$\sigma^2 = 2Dt_D \tag{2.7}$$

is helpful; D is the diffusion coefficient and t_D is the total time over which diffusion is occurring.

These basic, simple equations are the necessary tools for writing a quantitative statement of band spreading in terms of tangible, physical properties of a chromatographic system. We will now examine, and then develop, the quantitative arguments for each process, molecular diffusion, mass transfer, and "eddy diffusion," in turn.

2.5. MOLECULAR DIFFUSION

We are concerned here with diffusion of solute in the direction of flow. Diffusion coefficients of solute molecules in liquids are roughly five orders of magnitude lower than in gases. Thus we can see that while in gas chromatography band spreading due to longitudinal molecular diffusion is virtually entirely due to diffusion in the mobile phase, the situation is more complex when liquid mobile phases are used. In partition work significant diffusion in both stationary and mobile phases occurs (this also applies in gel permeation), while in adsorption chromatography one has to consider diffusion on surfaces as well as the obvious mobile phase effect. Giddings[5] has drawn attention to the need for the study of diffusion on surfaces. He considers that the rate of surface diffusion is likely to be broadly similar to rates of diffusion processes in liquids. It is likely, though unproven, that energy of adsorption and adsorbent geometry will control the rate of surface diffusion and thus influence the net band spreading. Presumably the more strongly a solute is adsorbed (i.e., the greater its heat of adsorption) the lower is its ability to diffuse along a surface. Since for a range of solutes of varying heats of adsorption the time spent adsorbed is roughly parallel to the heat of adsorption, it is reasonable to suppose that the total surface diffusive band spreading of these solutes is about equal. In the absence of factual information surface diffusion will be excluded from further consideration.

Molecules in liquids are held together by relatively strong intermolecular forces and the diffusion of a solute molecule will be impeded by such forces. Pictorially, the random-walk step length of a diffusing solute in a liquid will be about one molecular diameter. This should be compared to the equivalent situation in a gas, where the step length will be of the order of the mean free path (i.e., about 100 molecular diameters). Diffusion coefficients of relatively small solutes in similarly small solvents are typically around 10^{-5} cm^2 sec^{-1}. The actual value depends, naturally, on the magnitude of solute–solvent interactions and on the temperature.

We will now deduce the band spreading due to diffusion and then evaluate the factors influencing its extent.

Einstein's equation (2.7) enables us to calculate band spreading due to diffusion, $\sigma^2 = 2Dt_D$. The diffusion time t_D, is the time spent by the molecule in the appropriate phase, mobile or liquid. Considering diffusion in the mobile phase, in a chromatographic bed of length L through which the average mobile phase velocity is v,

$$t_D = t_m = L/v \tag{2.8}$$

since t_m is the time spent by the molecule in the moving phase. Thus $\sigma^2 = 2D_m L/v$, where D_m is the solute diffusion coefficient in the mobile phase. Also, from Eq. (2.6), $H = \sigma^2/L$; thus

$$H = 2D_m/v \tag{2.9}$$

However, in packed beds the distance a molecule actually diffuses is restricted through obstruction by the particulate matter and a correction factor (often called the "obstructive factor") γ_m is introduced. Hence, for mobile phase solute diffusion we have

$$H = 2\gamma_m D_m/v \tag{2.10}$$

In the stationary liquid phase of liquid partition systems, random solute diffusion can also occur. In this case if R_r is the fraction of the solute molecules in the mobile phase at any instant, then from (2.6) and (2.7)

$$H = 2\gamma_s D_s (1 - R_r)/R_r v \tag{2.11}$$

since $t_s = t_m (1 - R_r)/R_r$ and $t_m = 1/v$, where γ_s is the "obstructive factor" in the stationary phase, D_s is the diffusion coefficient of solute in the stationary phase, and v is the mean velocity of the mobile phase.

Equations (2.10) and (2.11) provide the required information on the contribution of longitudinal molecular diffusion, in the mobile and the stationary phases, respectively, to band broadening.

It is instructive to examine the magnitude of the factors involved. We have noted earlier that typical values of D_m are about 10^{-5} cm^2 sec^{-1}. Changing the mobile phase (i.e., the "solvent") can lead to changes in D_m of

Table 2.1. Viscosities at Various Temperatures (cP)

Substance	0°C	20°C	40°C	60°C
Diethyl ether	0.29	0.23	0.20	0.17
n-Hexane	0.40	0.32	0.26	0.22
Methyl acetate	0.48	0.38	0.31	—
Chloroform	0.70	0.56	0.47	0.39
Methanol	0.81	0.59	0.45	0.35
Benzene	0.90	0.65	0.59	0.39
Carbon tetrachloride	1.35	0.97	0.74	0.58
Isopropanol	4.56	2.37	1.33	0.80
Glycerol	4600	850	—	—

several orders of magnitude. An empirical expression, due to Wilke and Chang([5]), indicates that

$$D_m \propto TM^{1/2}/\eta \qquad (2.12)$$

where T is the absolute temperature, M is the molecular weight of the solvent, and η is the viscosity of the solvent. The viscosities of some typical liquid chromatographic mobile phases and, for comparison, those of glycerol are given in Table 2.1.

For the mobile phases listed the values of η^{-1} vary (at 20 °C) by about one order of magnitude, while M varies by a factor of only ~ 1.5. Thus viscosity is the dominant influence on D_m and, since one requires to minimize H, there is a superficial case for using the most viscous eluents at the lowest practical temperature. However, a typical value of γ is ~ 0.6 ([7]) and of ν is 1 mm sec^{-1} for "high-efficiency" work. Thus, from Eq. (2.10) we estimate that the contribution of diffusion in the mobile phase to overall plate height is about 10^{-3} cm. In comparison with typical plate heights for efficient liquid columns (i.e., about $1–5 \times 10^{-1}$ cm) this contribution is small. We can therefore conclude that there is little to be gained by optimizing the choice of mobile phase with respect to the diffusion of solutes in it; other factors are much more significant.

Regarding diffusion in the stationary phase, and considering liquid–liquid partition processes, the range of viscosities and molecular weights is much wider, reaching up to very viscous polymeric materials. Very approximately, assuming a stationary phase molecular weight 10,000 and viscosity ~ 1000 cP, diffusion is only one-hundredth of that in the mobile phase. Thus this factor is completely insignificant as a contributor to band broadening.

Although the influence of solvent on D_m may be quite large, we have seen that for typical materials the diffusion factor has only a marginal effect on column performance with even the best of presently available columns. The variation of D_m with change in solute is smaller than with change in solvent.

Frenkel[8] records that D_m in water changes from $\sim2 \times 10^{-5}$ cm^2 sec^{-1} for gases to $\sim0.3 \times 10^{-5}$ cm^2 sec^{-1} for sugar. The difference in chemical nature of these solutes is extreme but yields only an order of magnitude difference in D_m where, taking ether and glycerol as "extreme" examples of solvents, D_L for a given solute would vary by about three orders of magnitude [see data in Table 2.1 and Eq. (2.12)].

We conclude therefore that longitudinal molecular diffusion has a rather small influence on band broadening in liquid chromatography even with the most sophisticated of contemporary operating practices.

2.6. MASS TRANSFER

On a molecular level the passage of a band of solute through a chromatographic bed consists of a series of transfers of molecules back and forth between fixed and moving phases. In adsorption chromatography the random adsorption or desorption of molecules from the surface is occurring, while in a partition system solute molecules diffuse to the surface of the stationary liquid and may escape into the moving phase. In either system escape into the mobile phase requires that the solute molecules have a certain minimum energy. Transfer back into the stationary phase is a random process governed by molecular energetics and also by flow patterns which control the contacting of the phases.

For convenience we will term the process whereby a solute enters the stationary phase as "adsorption" and that when it leaves the phase as "desorption." Applying the random-walk concept, we note that a "desorption" step corresponds to a forward movement and an "adsorption" step to a backward movement with respect to the band center. In its passage through a bed, length L, a solute molecule makes a total number of steps twice the total number of adsorptions since desorption follows every adsorption. On average a molecule spends time t_m in the mobile phase between desorption and readsorption: during the time, it moves onward at the mobile phase velocity v. To cover the total bed length L thus requires a number n' of desorptions, given by

$$n' = L/vt_m \qquad (2.13)$$

and consequently, to twice this number of random walks $n = 2n'$,

$$n = 2L/vt_m \qquad (2.14)$$

We must here pause to remember that the spreading of chromatographic bands has to be considered in relation to the movement of the band center. The band moves only a fraction R_r of the movement of a molecule in the mobile phase, i.e., in time t_m the desorbed molecule moves vt_m and the band center moves $R_r v t_m$. Thus the relative movement of the molecule with respect to the center of the band is

$$L = (I - R_r)vt_m \qquad (2.15)$$

and this represents the actual average step length.

We know that $H = \sigma/L$, where L is the column length, and that $\sigma^2 = l^2 n$, where l is the average step length and n the number of steps.

Therefore, substitution of (2.14) and (2.15) into these expressions and eliminating σ gives

$$H = \sigma^2/l = L^2 n/l = [(1 - R_r)^2/l]v^2 t_m^2 \,(2l/vt_m)$$

i.e.,

$$H = 2\,(I - R_r)^2 v t_m \qquad (2.16)$$

This equation can also be written in terms of the mean time spent in the stationary phase t_s since

$$R_r = t_m/(t_m + t_s)$$

which can be written

$$R_r/\,(1 - R_r) = t_m/t_s \qquad (2.17)$$

and substituted back into (2.16) to give the relationship

$$H = 2R_r\,(1 - R_r)\,vt_s \qquad (2.18)$$

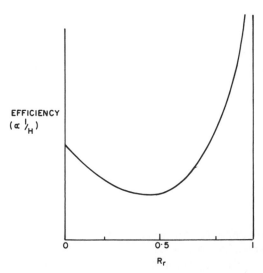

Fig. 2.7. Relation between efficiency and retention.

It can be seen that plate height increases linearly with mobile phase velocity. That is to say, the band spreading, or the distance a solute molecule lags or leads the band center, is greater the greater the rate of movement of of mobile phase. Equation (2.18) shows that a relationship exists between the chromatographic efficiency and the retention of a solute (Fig. 2.7), efficiency being relatively high for weakly retained materials, passing through a minimum for solutes moving at about half the mobile phase velocity, and then again increasing as retention further increases. At high retentions the undesirable consequences of large t_s values become prominent. As a rough generalization, therefore, highest efficiencies are found in systems where solute retentions and mobile phase velocities are low.

Looking at partition systems, the interchange of solute between the two phases approximates to the adsorption–desorption phenomena we have just considered. In each case a certain average time is required for the solute to be sorbed and desorbed; it is the underlying factors controlling these rates which are important. In partition what controls t_s is basically the time of diffusion t_D for a molecule to pass through a film of stationary liquid of thickness d. From Einstein's equation $\sigma^2 = 2D_s t_s$. For each sorption we can equate σ with d,

$$d^2 = 2D_s t_s \qquad (2.19)$$

and this can be substituted into the general mass transfer equation (2.18) to give

$$H = R_r(1 - R_r)vd^2/D_s \qquad (2.20)$$

for partition processes.

The linear dependence of H on mobile phase velocity naturally still holds; in addition we see that better efficiency (i.e., lower values of H) can be achieved by dispersing the stationary liquid so that its film thickness (or pool depth) d is as small as possible. One could crudely generalize that high efficiency in liquid–liquid chromatography is favored by use of relatively small quantities of stationary liquids of low viscosity [from Eq. (2.12)].

For a typical LLC separation

$$R_r(1 - R_r) \simeq 0.25$$
$$v \simeq 0.1 \text{ cm sec}^{-1}$$
$$d^2 \simeq 10^{-8} \text{ cm}^2$$
$$D_s \simeq 10^{-5} \text{ cm}^2 \text{ sec}^{-1}$$

Consequently, from (2.20) the plate height contribution of mass transfer in liquid partition chromatography is $\sim 2.5 \times 10^{-5}$ cm.

In adsorption chromatography the value of d, and also possibly D_s, differs from that in partition systems. d, the "depth" of the stationary phase, approximates to the depth of pores in the adsorbent and will inevitably vary with methods of preparation of the adsorbent and the particle size. As a crude approximation we can equate d with the particle radius, i.e., in high-efficiency systems it will be of the order 10^{-3} cm. D_s, the diffusion coefficient in the stationary phase, is a quantity even more difficult to estimate. It seems reasonable to suppose that this quantity will be somewhat lower than in stationary liquids deposited on porous supports; by comparison with the typical value of D_s quoted for partition systems, therefore, we will assume a value of 5×10^{-6} cm^2 sec^{-1} for adsorption chromatography. Since R_r $(1 - R_r)$ and v will typically have the same values in adsorption and partition, we conclude, by substituting the appropriate values into (2.20), that the plate height contribution of mass transfer in adsorption systems is 5×10^{-3} cm.

In exclusion chromatography, because the diffusion coefficient decreases as the molecular weight increases, so peak broadening also increases. For small molecules, peak broadening due to resistance to mass trasnfer is small compared with eddy diffusion, but for high-molecular-weight polymers its effect is significant.

Substituting the typical values $R_r(1 - R_r) = 0.25$, $d^2 = 10^{-5}$ cm^2, and $v = 0.1$ cm sec^{-1} into Eq. (2.20), we get

$$H \simeq 10^{-7}/D_s$$

Values of H calculated from this equation for several substances are shown in Table 2.2.

It should be noted that although the diffusion coefficients in the stationary phase are decreased by obstruction by the gel network, this effect seems to be small and has been neglected.

In summary, therefore, the mass transfer contribution to HETP is very small in partition chromatography ($\sim 10^{-5}$ cm), rather higher in adsorption ($\sim 10^{-3}$ cm), and becomes the dominant factor in exclusion chromatography of macromolecules ($\sim 10^{-1}$ cm).

Table 2.2. Mass Transfer HETP Contributions in Exclusion Chromatography

Substance	Diffusivity in toluene[9], cm^2 sec^{-1}	H, cm
n-Hexane	1.955×10^{-5}	8×10^{-3}
n-Hexatricontane	7.03×10^{-6}	2×10^{-3}
Polystyrene: mol. wt. 10,300	1.82×10^{-6}	9×10^{-2}
mol. wt. 97,200	5.13×10^{-7}	0.3

2.7. EDDY DIFFUSION

The flow of liquid through a packed bed is irregular and proceeds through a multiplicity of different paths which are interconnected but differ in their tortuosity and degree of constriction. Different flow paths vary in length and because of this and the local variations in flow velocity the time of transit through the packed bed along the flow paths varies considerably around the mean. This variation is a potent source of band spreading.

The so-called "classical" theory of eddy diffusion took account only of the flow stream patterns in a packed bed. Giddings in his "coupling" theory has drawn attention to the need to consider the possible diffusive movement of a solute from one stream path to another. He has distinguished five sources of variation in flow velocity through a particulate bed. These are as follows:

(1) The "transchannel" effect. In a flow channel, velocities are greatest along the axis and least close to the walls.

(2) The "transparticle" effect. In liquid partition chromatography the mobile phase will occupy that part of the pores of the solid support not filled by the stationary liquid, while in adsorption chromatography the pores will be largely full of mobile phase. In each case the entrapped mobile fluid will be essentially stagnant but surrounded by moving liquid.

(3) The "short-range interchannel" effect. Considerable velocity differences occur between adjacent open and constricted flow channels.

(4) The "long-range interchannel" effect. The short-range effects noted in (3) become even more significant over larger zones of a packed bed: mean velocities in adjacent zones (as opposed to adjacent flow channels) show marked variations.

(5) The "transcolumn" effect. On a column or bed wide-scale flow velocities at a containing wall may be as much as 10% greater, and at the center 10% less, than average. Additionally, bending or coiling a column may cause flow inhomogeneities.

These effects are largely unexplored experimentally, although it can be clearly seen that regularity of packing and of particle size and shape are advantageous in reducing variations in mobile phase flow velocity in a particulate bed. Eddy diffusion, as considered above, is a factor governed entirely by the properties of the stationary phase and its consequence is the broadening of solute zones being eluted through the bed by the mobile phase; some solute molecules get into faster-than-average, some into slower-than-average stream paths and accordingly lead or lag the band center. Each molecule will also pass from one flow path to another. For example, at the confluence of two paths the new, unified stream will have a velocity only vaguely related to those prior to the union. In addition, a solute molecule may diffuse from one flow path to another. For these reasons the passage of the molecule through the stationary phase bed consists of a series of erratic dashes or meanders.

Applying once more the random-walk approach to these processes, we note that in Eq. (2.7) σ may be set equal to d, where d is the distance a molecule diffuses in time t_e, the time needed for a molecule to transfer from one velocity regime to another. Depending on the precise mechanism of transfer (i.e., interchange governed by one or other of the factors discussed earlier), the diffusion distance d will vary. For the general case, therefore, we will equate d to $W_a d_p$, where the value of W_a depends on the process, and d_p is the particle diameter. (Use of d_p is justified because changes in flow velocities will occur roughly between each pair of adjacent particles because of the union of several streams at such points, as shown in Fig. 2.8. Bear in mind, however, that this two-dimensional representation is a poor reflection of the true situation.)

Thus from Eq. (2.7) we have

$$t_e = W_a^2 d_p^2 / 2D_m \tag{2.21}$$

The length of a step of the random walk, l, is the distance gained or lost on the mean position between changes from path to path. This distance is equal to $\Delta u \cdot t_e$, where Δu is the difference between the actual and the mean velocity $u \cdot \Delta u$ is some fraction W of the mean velocity (i.e., $\Delta u = uW$) and if we write $ut_e = S$, where S is the distance a molecule travels at a more or less constant velocity, then

$$l = ut_e = Wut_e$$
$$l = WS \tag{2.22}$$

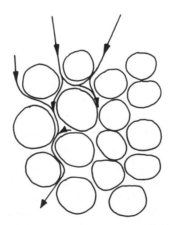

Fig. 2.8. Flow paths through a packed bed.

If the total migration distance is L, then the number of steps n is given by $n = L/S$; also, $l^2 n = \sigma^2$ and $H = \sigma/L$; hence

$$H = W^2 S \qquad (2.23)$$

since $S = u t_e$ and $t_e = W_a^2 d_p^2 / 2 D_m$ [Eq.(2.21)]. Then

$$H = W^2 W_a^2 d_p^2 u / 2 D_m \qquad (2.24)$$

W_a, it will be remembered, depends on the type of flow paths involved, while W_a is a fraction of the mean velocity, the magnitude of W also depending on the type of flow path. Thus we will combine W and W_a into one, flow-path-dependent, term W_i and rewrite (2.24) as

$$H = W_i d_p^2 u / 2 D_m \qquad (2.25)$$

Substituting typical values for d_p, u, and D_m of 10^{-3} cm, 10^{-1} cm sec^{-1}, and 10^{-5} cm^2 sec^{-1}, respectively, into Eq. (2.25), the order of magnitude of the "eddy diffusion" contribution to plate height is 10^{-2} cm.

2.8. FINAL EXPRESSION FOR PLATE HEIGHT

We are now in a position to gather together the various contributions to the total plate height in a chromatographic system, those of molecular diffusion in both the mobile and the stationary phases [Eqs. (2.10) and (2.11)], of mass transfer [Eqs. (2.18) and (2.20)], and of eddy diffusion [Eq. (2.25)]. Mobile phase velocity is seen to be a key factor in determining plate height and so the final expression can be written in the form

$$H = (B/u) + Cu + Au \qquad (2.26)$$

in which A, B, and C are short-hand coefficients having the physical characteristics we have already explored. This final equation summarizes the effect of processes occurring in a chromatographic bed on the shape (and in particular on the width) of a band of solute being washed through it. The equation has the form shown in Fig. 2.4, in which are indicated the approximate contributions of the individual terms (effects). Many gas chromatographers are very familiar with this type of graph and its proportions can be established easily for any column. The minimum plate height occurs at the "optimum mobile phase velocity"; in packed column gas chromatography, which has been studied extensively, this corresponds to volumetric gas flow rates of 1–50 ml min^{-1}, i.e., linear flow rates around 5 cm sec^{-1}. Under these conditions typical analysis times are in the range of 5–60 min. In liquid chromatography, as we have noted in developing the various plate height contributions, there are some most significant differences in the magnitude of effects compared with gas chromatography and these lead to a situation

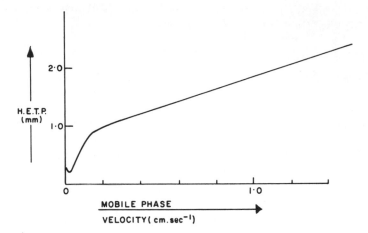

Fig. 2.9. Experimental HETP curve for an efficient column.
(Reproduced by courtesy of Elsevier Publishing Co.)

in which optimum mobile phase velocities are orders of magnitude lower.
An actual H–u curve for a liquid adsorption chromatographic column is
shown in Fig. 2.9[10] and it will be noted that the minimum H probably
occurs at a linear velocity of about 0.02 cm sec^{-1}, which is equivalent to a
volumetric flow rate of 0.1 ml min^{-1}. With this column typical analysis
times are to be reckoned in hours working at or near the optimum. Because
of the low mass transfer contribution, velocity can be increased without
increasing H too seriously in partition or adsorption. Thus much faster
analyses at high column efficiency can be obtained in practice.

Recapitulating on the plate height contributions of the several band
broadening processes, we concluded that they are of the following orders:

Molecular diffusion in the mobile phase	10^{-3} cm
Molecular diffusion in the stationary liquid phase	10^{-5} cm
Mass transfer contribution in partition	10^{-5} cm
Mass transfer contribution in adsorption	10^{-3} cm
Mass transfer contribution in exclusion	10^{-1} cm
"Eddy diffusion" contribution	10^{-2} cm

Allowing for the considerable approximations involved, it is nevertheless clear that the multipath diffusion term is a major contributor to band
spreading and that, since it involves the square of the mean particle diameter,
use of very fine particles is desirable. If the packing regularity can be improved, for example by using more symmetric particles or "better" packing
techniques, then a further decrease in H can be obtained. We see therefore
why so much current effort is devoted to improving column performance

through preparation of "tailor-made" supports or adsorbents of size down even to submicron diameter.

2.9. EXTRABED CONTRIBUTIONS TO BAND SPREADING

In this chapter we have been concerned primarily with the processes occurring in the bed of stationary phase as a solute is washed through it; however, the final concentration profile of the solute in the chromatogram as revealed by the detection system also includes the contribution of some other factors. The principal of these are: (i) the sharpness of the sample concentration profile at the start of the chromatography, (ii) nonlinearity of solute distribution between stationary and mobile phase as a function of concentration, and (iii) the possibility of further change in solute band shape after chromatography and before and during detection.

It is axiomatic that the wider the band of sample in the mobile phase introduced into a column or onto a thin layer, the wider will be the resolved component bands after the separation. Ideally, therefore, samples should be introduced as "infinitely narrow" charges to the chromatographic bed. Methods of sample introduction into columns and onto thin layers will be discussed in later chapters.

Nonlinearity of solute distribution between phases arises in both adsorption and partition processes and results in development of asymmetric peaks having displaced peak maxima (undesirable for qualitative analysis) and in loss of resolution. Generally, asymmetry arises where solute concentrations are too high or where very strong polar interactions between solutes and chromatographic phases occur.

The requirement that separated solutes must be detected after chromatographing dictates that further manipulation is necessary. For example, effluent from a column may be transported to, say, a continuous ultraviolet monitor, while a thin-layer chromatogram may be sprayed with a suitable revealing reagent. These operations can, and often do, introduce further peak broadening. Again this is essentially a matter of practical technique and we will give detailed consideration to it at the appropriate points.

2.10. COLUMNS VERSUS PLATES

The mechanisms we have been considering have been treated as occurring in packed beds, that is, either in columns or on relatively thin, flat layers. Essentially the form of the packed bed is immaterial to the treatment, but there are obvious differences between the two types of bed. The two important ways in which a thin layer differs from a column with respect to the mechanism of chromatography are its very different aspect ratio and its "openness."

A column has a very high aspect ratio; axial transport (accompanied by longitudinal diffusion) is great, whereas radial diffusion is limited by the constraining column walls. On a thin layer we have essentially a two-dimensional bed in which diffusion perpendicular to the direction of the relatively limited longitudinal transport of solute is not normally restricted. Consequently, band spreading occurs both longitudinally and "radially" on thin layers. The "openness" of thin layers has some valuable practical advantages, but some important disadvantages. Mechanistically, the serious limitation openness introduces is the inability to significantly change the mobile phase velocity, which is governed by such factors as the viscosity and surface tension of the eluent and by the temperature and the geometry of the chamber in which the chromatography is conducted. Additionally, the rate of movement of the eluent generally decreases as the distance from the starting point increases. Solutes are initially exposed to a rapidly moving flow of mobile phase, probably well above the optimum for good resolution. Thus the experimenter has little control over one of the most important variables in chromatographic processes.

The advantages and disadvantages of columns and of thin layers will emerge throughout this volume, but the principal, unsurmountable difficulty with TLC is the impossibility of controlling the mobile phase transport. It is nevertheless important to appreciate that for many practical purposes the efficiency of TLC separations is adequate and the other advantages of that technique so great that use of column chromatography is quite unjustified.

2.11. SUMMARY—IMPORTANT PRACTICAL CONSEQUENCES OF THEORY

Throughout the foregoing theoretical study of chromatographic mechanisms we have sought to highlight practical consequences. In this section we seek to gather together these conclusions in terms of the choice of materials and operating conditions open to the chromatographer. We do, however, introduce certain other essential factors (which are discussed in detail in later chapters) which inevitably modify these conclusions. Each of the following important parameters is thus considered in turn in the following pages: (1) the stationary phase: choice, particle size, length and thickness (diameter) of the bed. (2) The mobile phase: choice, velocity. (3) Operating temperature. (4) Availability of appropriate instrumentation.

2.11.1. Choice of Stationary Phase

The stationary phase is generally chosen on the basis of the selectivity required to provide adequate separation of the sample. It has to be considered in conjunction with the choice of mobile phase and any constraints imposed by other operating parameters. Secondary considerations which might be involved in certain circumstances include the capacity of the phase, if relatively large-scale separations are needed, and the purity (or reproducibility of phase properties) if reproducible retention data are sought.

The selectivity of stationary phases is considered in the next chapters; a wide range of properties can be used to provide the necessary separation Molecular size, polarity, hydrophilic–lipophilic balance, molecular shape, presence of specific functional groups, all these facets of molecular structure can be turned to advantage in choosing the most appropriate phase for a particular job. We will not anticipate succeeding chapters here, but emphasize the need to think carefully about structural features of sample components which could provide the bases of achieving the desired separations.

2.11.2. Stationary Phase Particle Size

Theory predicts that particle size will affect efficiency of chromatographic separations; practice not only confirms this, but also exposes other side effects of particle size which have to be taken into account.

The theories we have mentioned predict that HETP increases with increasing particle size [Eqs. (2.1) and (2.26)]. Because of the further complication of particles actually covering a size range, this effect increases as the size range widens. Packing irregularities are at a minimum with uniform sized particles. They increase progressively as the size distribution broadens, not only because of the intermeshing of the different sized particles, but also because of the tendency for nonuniform distribution of the different sizes throughout a bed.

A further theoretical prediction is that in adsorption and exclusion chromatography the mass transfer term critically depends on the depth a solute can penetrate into the stationary phase—the "film thickness" of partition systems approximates to the particle diameter in adsorption and exclusion.

There are practical problems superimposed on these factors, such as the ability effectively to pack columns with fine particulate matter. Also, the ease of adhesion of thin layers is a function of particle size. Further, size affects the permeability and therefore the pneumatics of the system and the speed of analysis.

The conclusion from theory is that infinitely small, uniform-sized particles are preferred. Practice modifies this as follows. In thin-layer work very fine particles (1–10 μm) are ideal since layers of these particles bind rather easily to the supporting plate and undesirable additives need not be used to improve binding. In column chromatography the difficulty of consolidating beds of packing of mean diameter less than 20 μm seems to define the present limit to which the theoretical ideal can be regularly approached. It is unlikely this limitation will remain for long and some years ago highly efficient columns packed with submicron particles were described([11]). This is in fact an area (column packing) in which steady progress can be anticipated. The mass transfer contribution to plate height can already be reduced markedly by use of controlled-surface-porosity or surface-coated materials described elsewhere. In summary, for column chromatography the present recommendation would be to use particles of a mean diameter between 20 and 30 μm and a range preferably not exceeding about 15 μm; alternatively, the surface-coated adsorbents and supports can be used to good effect.

2.11.3. Dimensions of the Stationary Phase Bed

Dismissing thin-layer chromatography rather briefly, one notes that the bed length is restricted in practice to about 20–30 cm by the limited ability of mobile phases to move over greater distances in acceptable times and also by increasing manipulative inconvenience as plate size increases. The bed (layer) thickness for optimum work, reflecting a balance between efficiency and sample capacity, is of the order of 100 particle diameters and is normally set at about 250 μm except for "preparative scale" work, where layers are often up to 1000 μm thick.

In column chromatography the situation is more complex and inevitably decisions are a compromise between several conflicting requirements. The range of column diameters which can be recommended for high-efficiency work is 1–10 mm; the difficulty of packing increases as the size diminishes and the column efficiency falls off as it increases, setting these limits. A good choice would be 4-mm-i.d. columns, identical, in fact, with those still most commonly used in gas chromatography. The length of columns used very much depends upon the number of theoretical plates required and the ability to force mobile phase through the column at a fast enough rate to give an acceptable analysis time. At present the best practical advice is to pack columns of length not exceeding 1 m (this is about the limit for adequate consolidition of stationary phase to yield plate heights of the order of 1 mm at flow rates not too far from the optimum). If resolution of such a column is inadequate, further 1 m lengths should be added, making intercolumn links with minimal volume capillary connections([10]). Each additional length of column decreases the permeability and thus, to maintain acceptably small analysis times, higher driving pressures are needed. For 4-mm-i.d. columns 1 m in length, packed with adsorbents in the size range around 40 μm, the pressure required to give a flow rate of 1 ml min^{-1} is about 20 atm. One can thus deduce the approximate limiting column length usable, bearing in mind the available driving pressure and the analysis time sought. Most work in high-performance liquid chromatography has so far been with columns not exceeding about 5 m length.

2.11.4. Choice of Mobile Phase

Various physical and chemical properties govern this choice. The most important factor is the influence of the mobile phase on the selectivity of the system; solubility of samples and influence of such properties as surface tension and viscosity are less important.

The role of the mobile phase with regard to selectivity is discussed elsewhere in this book; it must always be given first consideration. It is worth noting that in adsorption chromatography the use of mobile phases of changing composition (increasing eluent strength) can often considerably speed up a separation.

Solubility of some samples, especially polymers, limits the choice of the mobile phase and the use of certain detectors imposes constraints, as will be discussed in detail later. An obvious but by no means unique detector-imposed constraint is where ultraviolet spectrophotometric detection is used; in such a case many common solvents, including, of course, all aromatic substances, are unsuitable. Another consideration in systems where long analysis times and automatic or unattended running is involved is the safety aspect flammability and toxicity have to be borne in mind.

The dynamics of chromatography we have discussed are of relatively minor importance in selecting a suitable mobile phase. Generally, viscous fluids are undesirable because they require higher driving pressures than the normal eluents for equal analysis times. The influence of physical properties of the mobile phase on peak broadening is so small as to be insignificant compared with the other factors summarized above.

2.11.5. Mobile Phase Velocity

On this subject we have seen that theory has much to tell us. There is an optimum flow rate at which the conflicting effects of molecular diffusion (minimized by high flows) and resistance to mass transfer (minimized by low flows) are in balance. The low value of molecular diffusion coefficients in liquids leads to the conclusion, borne out by practical experience, that optimum linear velocities will be low. A typical value is 10^{-1} cm sec^{-1} for 4-mm-i.d. columns, which corresponds roughly to a volumetric flow of 10^{-1} ml min^{-1}. It is fortunate that in liquid chromatography HETP increases relatively slowly at velocities above the optimum[10], so that even at about 1–10 cm sec^{-1} plate heights are only a factor of 2–10 times that at the optimum. Consequently the inherently slow process of high-resolution liquid chromatography can be accelerated with significantly less loss of resolution than in gas chromatographic separations.

2.11.6. Operating Temperature

Distribution coefficients are temperature-dependent, so that some separations may be significantly improved by changing temperature. More subtle effects have been shown to occur in some adsorbtion systems[11] which can also provide improved selectivity. In terms of column performance, temperature plays only an indirect role, through its effect on properties such as viscosity and diffusion coefficients. Indeed, most liquid chromatography is carried out at room temperature and the principal advantage to be gained by use of higher temperatures is probably increased speed of analysis, though at the same time separation factors tend to be reduced. In adsorption chromatography faster analyses can more effectively be achieved by progressively increasing the eluent strength.

2.11.7. Availability of Appropriate Instrumentation

Instrumental factors which affect the ease of attainment of adequate separation in an acceptable time are principally (i) the length and diameter of column which can physically be accommodated and with which the pumping system can cope and (ii) the design of injection and detection systems to minimise extracolumn band spreading. At the time of writing, instrument manufacturers have clearly recognized these factors and are marketing increasingly successful units capable of providing excellent performance when column packings and mobile phase composition and velocity are properly chosen.

REFERENCES

1. Giddings, J. C., *Dynamics of Chromatography,* Part I, Marcel Dekker, New York, 1965.
2. Wilson, J. N., *J. Am. Chem. Soc.* **62**, 1583 (1940).
3. Martin, A. J. P., and Synge, R. M., *Biochem. J.* **35**, 1358 (1941).
4. Van Deemter, J. J., Zuiderweg, F. J., and Klinkenberg, A., *Chem. Eng. Sci.* **5**, 271 (1956).
5. Ref. 1, pp. 242–243.
6. Wilke, C. R., and Chang, P., *Am. Inst. Chem. Eng. J.* **1**, 264 (1955).
7. Ref. 1, p. 35.
8. Frenkel, J., *Kinetic Theory of Liquids*, Oxford Univ. Press Chapter 4, (1946).
9. Kelley, R. N., and Billmeyer, F. W., Jr., *Anal. Chem.* **41**, 874 (1969).
10. Stewart, H. N. M., Amos, R., and Perry, S. G., *J. Chromatog.* **38**, 209 (1968).
11. Maggs, R. J., and Young, T. E., *Gas Chromatography 1968,* Harbourn, C. L. A., ed., Inst. Petroleum, London (1969).

FURTHER READING

Knox, J. H., and Saleem, H., *J. Chromatog. Sci.* **7,** 614 (1969). (Discusses kinetic conditions for optimum speed and resolution in column chromatography.)

Smuts, T. W., van Niekerk, F. A., and Pretorius, V., *J. Gas Chromatog.* **5,** 190 (1967). (Discusses the effect of operating parameters on speed of liquid chromatography.)

Huber, J. F. K., in *Physical Separation Methods,* Vol. IIB, Wilson, C. L., and Wilson, D. W., eds., Elsevier, Amsterdam Chapter 1, (1968). (Contains a presentation of the theory of liquid chromatography in columns.)

Chapter 3

Adsorption Chromatography: Mechanism and Materials

3.1. INTRODUCTION

Adsorption may be defined as the concentration of solute molecules at the interface of two immiscible phases. In liquid–solid adsorption chromatography (LSAC) the mobile phase is a liquid while the stationary phase is a finely divided, usually porous solid. The atoms in the bulk of the solid are subjected to equal forces in all directions, whereas the surface atoms experience unbalanced forces which can attract molecules from the surrounding solution to restore the balance.

In a multicomponent system selective adsorption occurs due to competition between the solutes and the mobile phase for the surface. It is governed by the differences in the strengths of the adsorption forces between the adsorbent and the adsorbates. In general, polar compounds are more strongly adsorbed by polar solids than are nonpolar compounds. Adsorption of a polar compound is enhanced in a nonpolar medium, but reduced in a polar medium, due to increased competition of the mobile phase for the surface.

A convenient method of studying adsorption is by measurement of adsorption isotherms.

3.2. ADSORPTION ISOTHERMS

An adsorption isotherm describes the equilibrium concentration relationship between the adsorbed and unadsorbed solute at a given temperature. It is a plot of the concentration of solute in the adsorbed phase versus

CONCENTRATION
OF SOLUTE IN
ADSORBED PHASE

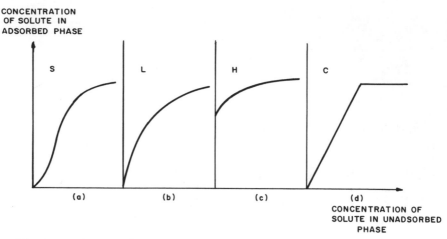

CONCENTRATION OF
SOLUTE IN UNADSORBED
PHASE

Fig. 3.1. Basic types of adsorption isotherms between a liquid and a solid surface.

its concentration in the unadsorbed phase. Giles *et al.*([1]) classified the adsorption of solute from a solvent onto a solid surface according to the following scheme (see Fig. 3.1).

1. The *S*-shape isotherm represents the situation in which, as adsorption proceeds, it becomes easier for the solute molecules to be adsorbed; those already adsorbed on the surface at the most active sites assist further adsorption by intermolecular bonding. It is found that such isotherms are generally, but not always, given by flat molecules standing edge-on to the adsorbent surface, e.g., phenol adsorbed on alumina, where the hydroxyl group probably forms a hydrogen bond to surface oxygen atoms on the alumina, and the aromatic nucleus associates with other solute molecules.

2. The *L* (or normal Langmuir) isotherm is the most common one met with in LSAC. As adsorption proceeds, the most active sites are first covered by adsorbate and the ease with which adsorption takes place decreases until finally the monolayer is complete, and all the adsorption sites are occupied. This type of isotherm is usually obtained when molecules are adsorbed flat and when there is no intermolecular bonding.

3. The *H* (high-affinity) isotherm starts at a positive value on the ordinate axis, showing that all the solute has been removed from dilute solution. This isotherm is typical of chemisorption.

4. The *C* (constant-partition) isotherm is linear. This indicates that, as adsorption proceeds, the ease with which it takes place remains constant. This type of isotherm, though common in partition chromatography, is rarely observed in LSAC.

Adsorption isotherms can be further subdivided according to subsequent inflections and plateaus. Usually, a further rise following an initial plateau, or merely an inflection, indicates the formation of a second layer on top of the first, or in some cases, a reorientation of the first layer.

Maxima have been observed in some isotherms, mainly in the L and H classes. The solutes involved associate in solution and it has been suggested that at high concentrations the association may draw some of the adsorbed solute back into solution.

Further discussion of nonlinear isotherms and their practical significance follows in the section concerned with sample size in adsorption chromatography.

3.3. NATURE OF ADSORPTION FORCES

The forces involved in adsorption chromatography can be classified as follows.

3.3.1. Van der Waals Forces (London Dispersion Forces)

These are intramolecular forces which hold neutral molecules together in the liquid or solid state. They are purely physical in character and do not involve the formation of any chemical bonds. Adsorption of this type is known as physical adsorption, and is characterized by low adsorption energies leading to the rapid establishment of equilibria and hence good chromatographic separation.

Dispersion forces account for all the adsorption energy in cases of adsorption of nonpolar solutes onto nonpolar adsorbents, e.g., hydrocarbons on graphite, while Snyder has shown[9] that the contribution of dispersion forces on alumina ranges from 100% for saturated hydrocarbons to less than 50% for polar molecules such as acetone or methanol.

3.3.2. Inductive Forces

These exist when a chemical bond has a permanent electrical field associated with it, e.g., a C–Cl or C–NO_2. Under the influence of this field, the electrons of an adjacent atom, group, or molecule are polarized so as to give an induced dipole moment. It seems[2] that induction forces make a major contribution to the total adsorption energy on alumina (but not on silica).

3.3.3. Hydrogen Bonding

These forces make an important contribution to the adsorption energy between solutes having a proton-donor group and a nucleophilic polar surface possessed by, for example, silica or alumina, which is normally covered with hydroxyl groups. Similarly, the hydroxyl groups on the surface may react with other weakly electrophilic groups such as ethers, nitriles, or aromatic hydrocarbons.

3.3.4. Charge Transfer

This could occur, for example, when an electron is transferred from a solute S to a surface site A to form an adsorbed complex of the type S^+A^-. However, it has been shown[2] that forces due to charge transfer make an insignificant contribution to the adsorption energy of most compounds.

3.3.5. Covalent Bonding (Chemisorption)

This occurs when chemical bonds are formed between solute and adsorbent. These relatively strong chemical forces give rise to H-type isotherms and generally lead to poor separation in elution chromatography. Chemisorption is often exploited for the selective retention of certain compound types, e.g., the adsorption of amines by cation-exchange resins, the adsorption of olefins by silver nitrate-impregnated silica. On the other hand, H-type isotherms are not uncommon in high-efficiency elution chromatography. They can be attributed to the chemisorption of certain solutes onto those active sites on the surface of the adsorbent that have not been fully deactivated. For example, silica surfaces may contain some residual acidic sites which chemisorb bases. Similarly, alumina contains basic sites which strongly chemisorb acids. Florisil (magnesium silicate) also contains strong acidic sites and has been observed to chemisorb a wide variety of compounds, including aromatic hydrocarbons, basic nitrogen compounds, and esters, while magnesia chemisorbs polynuclear aromatic hydrocarbons.

As a consequence of chemisorption in columns, certain solutes give rise to strongly tailing elution bands, resulting in incomplete resolution and sample recovery, while in TLC part of the sample is seen to remain behind as a spot at the point of application.

For a further discussion of chemisorption see list of further reading at the end of this chapter.

3.4. CHOICE OF CHROMATOGRAPHIC SYSTEM IN LSAC

In designing a suitable chromatographic experiment in LSAC, the following factors have to be taken into consideration: (i) choice of adsorbent (ii) sample size and linear capacity, (iii) standardization of adsorbent, (iv) choice of mobile phase, (v) method of detection or quantitation.

The last point, method of detection or quantitation, will be dealt with in the respective chapters on thin-layer and column chromatography, although in column chromatography especially the choice of a particular mobile phase may be influenced by the detection system used. In addition, the method of detection determines the sample size required. Each of the other four factors will now be discussed in turn.

3.4.1. Choice of Adsorbent

(a) General Properties of Adsorbents

Adsorbents used for LSAC are finely divided, porous solids with a surface area usually greater than 50 m^2 g^{-1}. Table 3.1 shows the more common materials used as adsorbents, placed approximately in order of increasing strength. Strong adsorbents, i.e., those adsorbents with a relatively high concentration of strongly active sites, are preferred for the separation of weakly adsorbed, chemically inert compounds such as hydrocarbons, while weak adsorbents are preferred for the separation of labile or strongly adsorbed compounds.

Although the nature of the adsorbent surfaces has been indicated, this is frequently modified by the presence of free acid or base left over from the manufacturing stage or by the deliberate addition of buffering agents.

Other materials that have been used as adsorbents include calcium sulfate, talc, polyamide, organo-clays, and molecular sieves. A number of group-selective adsorbents have been prepared by impregnating an adsorbent with a material that will form a complex with a specific organic functional group. For example, silver nitrate-impregnated silica gel has been used for the separation of unsaturates. Further examples will be discussed in the section on modified adsorbents.

Silica gel and alumina are by far the two most common adsorbents in use today. It is not now necessary for one to prepare one's own adsorbent since they are readily available commercially. In fact, one can go even further,

Table 3.1. Activated Adsorbents in Approximate Order of Increasing "Strength"

Adsorbent	Nature of active sites
Sucrose	Neutral
Starch	Neutral
Kieselguhr	Neutral
Silica	Acidic
Magnesium silicate	Acidic
Alumina	Acidic and basic sites
Fuller's earth	Acidic
Magnesia	Basic
Charcoal	Neutral and acidic
Ion-exchange resins	Acid or basic species

Table 3.2. **Some Commercial Sources of Adsorbents** [a]

Chemical name	Trade name	Supplier
Magnesium silicate	—	Hopkin & Williams, Ltd. Chadwell Heath Essex England
	—	M. Woelm, D-344 Eschwege, W. Germany
	Florisil	Floridin Co., Pittsburgh, Pa. 15235
Magnesia	—	Hopkin & Williams, Ltd. Chadwell Heath Essex England
Carbon	Graphon; Spheron	Cabot Corp. Cambridge, Mass.
Ion exchangers	Amberlyst	Rohm & Haas, Co., Independence Mall West Philadelphia, Pa. 19105
Surface-modified glass beads	Zipax; Permaphase	E.I. Du Pont de Nemours & Co. Wilmington, Del. 19898
	Corasil	Waters Associates, Inc. 61 Fountain Street Framingham, Mass. 01701.
Textured glass beads	Corning glass beads	Corning Glass Works, Corning, N. Y. 13840

[a] Excluding silica and alumina.

since precoated TLC plates are now widely used and prepacked, high-efficiency columns are gradually being introduced. For a list of suppliers of chromatographic silica and alumina, see Heftmann[35]. Table 3.2 lists some suppliers of other adsorbent materials referred to in this chapter.

In selecting a suitable adsorbent, one needs to consider "adsorbent type" (i.e., strength, polar or nonpolar, surface pH) and the surface area and pore diameter. These factors will now be considered in turn.

(b) Adsorbent Type

The various adsorbent types exhibit different selectivities toward different compound types. Polar adsorbents (metal oxides, magnesium silicate, etc.) selectively adsorb unsaturated, aromatic, and polar molecules such as alcohols, amines, and acids. Polar adsorbents may be further subclassified as acidic, basic, or neutral, according to the pH of the surface. Silica, magnesium silicate, and cation-exchange resins are acidic and thus chemisorb bases. While chemisorption is an effective concentration method, quantitative chromatographic separation may not be possible because of the difficulty of desorption. Bases are best separated on basic adsorbents such as magnesia. Similarly, basic adsorbents chemisorb acids and these are best separated on acidic adsorbents. The alumina surface contains both acidic and basic sites, but this is an excellent adsorbent for unsaturated and aromatic compounds.

Nonpolar adsorbents such as graphitized carbon, which is a strong adsorbent, and kieselguhr, which is a weak adsorbent, show no selectivity for the adsorption of polar molecules. Kieselguhr is such a weak adsorbent that it has been used to provide an inactive solid support for the stationary phase in liquid partition chromatography (see Chapter 4). Further details on the nature of adsorbent surfaces are discussed in the section on individual adsorbents.

(c) Surface Area and Pore Diameter

The surface area and pore diameter of a given adsorbent vary widely with the method of manufacture. Probably no two commercial manufacturers produce silicas of the same surface area and pore diameter. The variation in properties of different batches of the same grade of adsorbent from one manufacturer is usually not great. Dramatic changes do occasionally occur, however, presumably due to alterations in process conditions. Attention must be drawn to those instances where the same manufacturer operates plants in more than one location. The properties of the adsorbents from the alternative locations are rarely identical. As a result, the reader is recommended to pay particular attention to the properties of his adsorbent and to standardize on one particular grade where possible.

Adsorbents for chromatography are porous solids with high specific surface areas usually in excess of 50 m^2 g^{-1} to provide high sample capacity. The surface area increases as the porosity increases and the average pore diameter decreases. The linear adsorption coefficient K° of a solute is independent of both these parameters provided the solute molecule is small enough to enter the pores unimpeded and provided the nature of the active surface sites is independent of the pore diameter. We shall see that in the case of silica gel, for example, large differences in pore diameter correspond to differences in surface structure. If the size of the solute molecule is comparable with or exceeds the pore diameter, then the phenomenon of exclusion rather than adsorption occurs. Exclusion is discussed fully in Chapter 5.

3.4.2. Sample Size and Linear Capacity

(a) Sample Size and Band Shape

In analytical LSAC relatively low sample concentrations are used, which means only the low-concentration regions of the isotherms need be considered. In these regions the isotherms have three possible shapes: convex (*L*-type isotherm), concave (*S*-type isotherm), and linear (*C*-type isotherm). These three types give rise to elution bands or spot shapes as shown in Fig. 3.2.

Thus, in the case of convex isotherms, which are the normal ones encountered in LSAC, the adsorption coefficient, K decreases as the sample size increases. Only at very low sample loadings can the isotherm be considered to be approximately linear. The consequence of this is illustrated in Fig. 3.3, which shows the effects of increasing sample loadings upon the band shape, in column chromatography, and the retention value, in thin-layer chromatography, as is observed in practice.

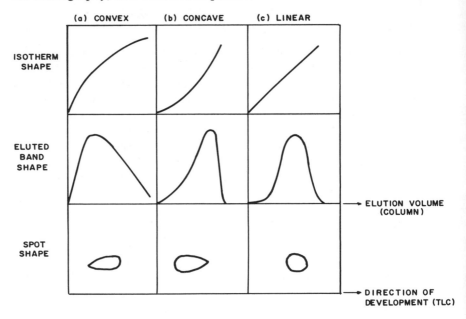

Fig. 3.2. Effect of isotherm shape on eluted band or spot shape.

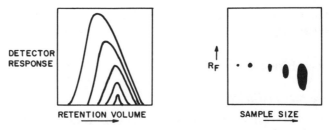

Fig. 3.3. Effect of sample size on band shape and retention values.

At very low sample loadings, V_r and R_F apparently remain constant. As the sample load is increased, however, a point is reached where V_r and R_F noticeably decrease. Increasing the sample size still further eventually leads to an increase in the amount of band tailing, loss of resolution, and even incomplete elution from the bed.

(b) Isotherm Linearity

On the theoretical side, assumption of isotherm linearity is essential to the development of any general theory of adsorption chromatography. From such a theory understanding of behavior in nonideal separations follows. On the practical side, a linear isotherm means that R_F, V_r, and K are constant for a given system at a given temperature, i.e., they are independent of concentration, and identification of separated zones by retention values is facilitated. Furthermore, linear isotherms give rise to Gaussian elution peaks or symmetric spots, enabling band resolution to be optimized.

Prior to the 1950's it was assumed that adsorption isotherms in LSAC were generally nonlinear. This was perfectly true with the large sample-to-adsorbent ratios which were and still often are being used. The development of TLC during the 1950's demonstrated that under the conditions in which this technique was used not only were faster and sharper separations obtained, but symmetric spots were common—suggesting that isotherm linearity was readily attainable using small sample loadings on deactivated adsorbents. Only during the past few years have the practical advantages of linear isotherm separations in columns been fully exploited.

(c) Linear Capacity

The linear capacity of an adsorbent has been defined([3]) as the maximum weight of sample that can be applied to a gram of absorbent before the adsorption coefficient K falls more than 10% below its linear isotherm value $K°$. This means that as long as the sample size does not exceed this value, the V_r or R_F of the individual solutes will be within 10% of their linear isotherm values.

The linear capacity of most activated adsorbents is very low (less than 10^{-4} g g^{-1}). It is therefore desirable to find ways of increasing the linear capacity to aid in the detection and further handling of sample components. Before we do this, however, let us look at some of the reasons why isotherms are nonlinear in adsorption chromatography.

(d) Origins of Isotherm Nonlinearity

Isotherm nonlinearity arises from three main causes([2]):

Surface coverage. It can be shown from the theory of the Langmuir isotherm that this becomes nonlinear when 10% of the adsorbent surface is covered by sample. This fixes the maximum possible linear capacity of a given adsorbent for a given solute. It can be calculated that about 0.03 g of a typical organic compound would form a monolayer on 100 m² of surface. Thus the maximum linear capacity of an adsorbent with a surface area of 100 m² g^{-1} would be 0.003 g g^{-1}. In practice, however, much lower linear capacities are generally encountered.

Interaction between adjacent adsorbed sample molecules. Such interactions are insignificant at the low surface coverage imposed by the above restriction and therefore need not be considered further.

Adsorbent heterogeneity. This arises because not all adsorption sites on a surface are equivalent. Some sites are said to be more "active" than others. The greater the heterogeneity, the smaller the linear capacity becomes. A small number of very active sites greatly reduces the linear capacity because these sites are occupied by solute molecules first. Departure from linearity occurs when 10% of these active sites have been covered. Heterogeneity is therefore seen to be the major cause of low linear capacities and ways must be sought to overcome this.

(e) Maximization of Linear Capacity

The more active sites of a heterogeneous surface can be deactivated by the addition of a polar solvent such as water, thus increasing the linear

Table 3.3. Retention Volume and Linear Capacity Data for Elution of Naphthalene by Pentane from Various Silicas Initially Activated at 195°C [a]

Pore size	Silica	Starting surface area, m² g^{-1}	Water added, wt%	V_r, ml g^{-1}	Linear, g g^{-1} × 10^{-4}	Capacity, g m^{-2} × 10^{-7}
Narrow	Davison Code 12	801	0	51.9	0.9	1.1
			2.0	25.1	7.5	
			7.5	6.3	29	
			16.0	1.84	26	
Medium	Davison MS	866	0	46.4	1.0	1.2
			2.1	18.0	2.5	
			7.9	5.0	15	
			16.8	2.16	25	
Wide	Davison Code 62	313	0	6.5	1.9	6.0
			0.8	4.42	2.9	
			2.9	2.74	11	
			6.2	1.77	4.2	

[a] Reprinted from Ref. 4 by courtesy of Marcel Dekker, Inc., New York, and L. R. Snyder.

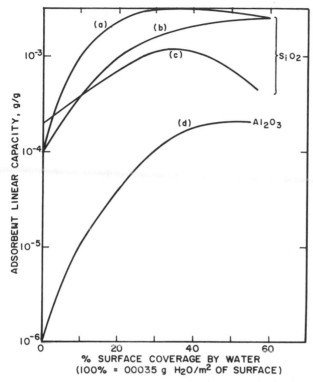

Fig. 3.4. Adsorbent linear capacity as a function of adsorbent type and relative deactivation by water. [Reproduced by courtesy of Marcel Dekker, Inc., and L. R. Snyder[2].]

capacity. Table 3.3 shows the effect of deactivation by water on the retention volume and the linear capacity of three silica gels differing in porosity and surface area.

The results are illustrated graphically in Fig. 3.4, which also includes data for a medium-pore alumina (Alcoa F-20) initially activated at 400°C. Wide-pore silica (Fig. 3.4c) has a relatively uniform surface and a low surface area. Since heterogeneity is a more dominant factor than surface area, the linear capacity of the activated adsorbent is higher than that of the medium-pore (Fig. 3.4b) or narrow-pore (Fig. 3.4a) silica. Because so few active sites are present in the wide-pore silica, the maximum linear capacity is reached after the addition of a relatively small amount of water. Addition of further water results in the coverage of the remaining uniform surface, resulting in a decrease of linear capacity. The narrow-pore silica, on the other hand, possesses a large surface area and a relatively high surface heterogeneity. Hence, the linear capacity of the activated adsorbent is low. A larger amount of wa*er is required to deactivate all the active sites, but when full deactivation has been achieved the linear capacity is greater than that of the wide-pore silica because of its greater surface area.

In general, we can say that the addition of 1–2% water per 100 m^2 g^{-1} of surface of polar adsorbent such as alumina, silica, and other metal oxides (corresponding to 30–60% surface coverage) increases the linear capacity 5–100 fold. Where highly polar eluents or elevated temperatures are to be used, physically adsorbed water is likely to be lost, resulting in some reactivation and a reduction in linear capacity. In these circumstances deactivation is best achieved using glycol or glycerol[5]. Deactivation of adsorbents such as charcoal, which does not adsorb water, can be achieved with certain high-molecular-weight organic compounds such as cetyl alcohol or stearic acid[6].

The linear capacity of a heterogeneous adsorbent can also be increased by decreasing the value of the adsorption coefficient K. This can be achieved by: (i) increasing the temperature; however, polar adsorbents tend to lose adsorbed water on heating and become more activated, i.e., the surface would become even more heterogeneous; or (ii) increasing the polarity of the solvent; however, if K is reduced too much by this means, resolution is lost.

In summary, it is seen that the most useful way of increasing the capacity of a heterogeneous surface is to deactivate the active sites by the addition of a polar liquid.

(f) Linear Capacities in Columns and Thin Layers

In a column the solute band is in contact with only a fraction of the total adsorbent in the column at any instant. This means that the linear capacity of the column is related to the width of the solute band[2]. In an efficient column, solute bands are narrow and therefore the linear capacity is apparently low. But since efficient columns are generally used for analyzing complex mixtures, as the separation proceeds more solute bands are in contact with adsorbent at any instant. Therefore the linear capacity of an efficient column increases with increasing complexity of the mixture. Those components that are not adsorbed at all can be disregarded in calculating the linear capacity of the column.

In analytical scale TLC the sensitivity of the various detection methods used often enables less than 1 μg of a sample component to be located. The usual practice is to apply sample loadings within the range 10–50 μg to adsorbent layers of 250 μm thickness, which is the optimum layer thickness for maximum speed and resolution. mpleSa loadings within this range are well below the maximum linear capacity (75 μg), assuming it to be applied as a spot to 1 cm^2 silica gel, 250 μm thickness on a plate, and the chromatogram developed in the norm lamanner (see Chapter 6).

3.4.3. Adsorbent Standardization

The value of $K°$ of a particular solute in a given mobile phase–adsorbent system depends upon the surface area and surface activity of the adsorbent. Variations arise because of differing manufacturing processes and subsequent thermal treatment and deactivation of the final product. It is highly desirable to be able to obtain repeatable and reproducible $K°$ values so that experiments can be duplicated and so that R_F and V_r values can be accurately measured and used for identification purposes. The following equation shows how the adsorbent surface area and surface energy affect the value of $K°$[7]:

$$\log K° = \log V_a + \alpha(S° - A_s\varepsilon°) \tag{3.1}$$

where V_a is the adsorbent surface volume, equal to the volume of a solvent monolayer (approximately 0.00035 times the surface area), α is an adsorbent energy function, proportional to the average surface energy of the adsorbent, and the parameter $(S° - A_s\varepsilon°)$ is a constant for a particular combination of solute, eluent, and adsorbent type. This equation has been experimentally verified for a number of systems[7].

Values of V_a and α for some of the common adsorbents deactivated by adding varying amounts of water have been tabulated[7]. The results show that V_a and α decrease regularly as the water content increases. Families of curves (Fig. 3.5) have been constructed relating V_r or R_F versus water content, each curve corresponding to a given value of $(S° - A_s\varepsilon°)$. It can be seen that decreasing the activity of the adsorbent increases the mobility of the solute.

To prepare a standardized adsorbent it is safest to assume that the commercial adsorbent, as received, has an unknown water content. In TLC, standardization of the plate is achieved by equilibrating it in a tank containing an atmosphere of known relative humidity. Details on how this is done are described in Chapter 6. In column chromatography the adsorbent is first heated in air for periods of 4–16 hr at some specified temperature to remove

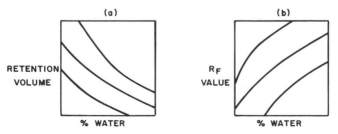

Fig. 3.5. Effect of water deactivation on (a) retention volume and (b)R_F value.

adsorbed water (see the section on individual adsorbents). This produces an adsorbent of definite (even if unknown) activity and water content. Deactivation is then carried out by the addition of a measured volume of water (or other deactivator) to a weighed amount of activated adsorbent contained in a stoppered flask. After a short period of vigorous shaking to disperse lumps the adsorbent is left to stand for 24 hr with occasional shaking, during which time the added deactivator distributes itself evenly over the whole adsorbent surface. Adsorbents prepared in this manner are stable indefinitely provided they are stored in tightly closed containers.

After deactivation the activity of the adsorbent should be checked by a standard procedure. In TLC a test mixture of the three dyes indophenol blue, p-dimethylaminoazobenzene, and Sudan red G (50 mg of each dissolved in 50 ml benzene) is developed in benzene using the standard procedure given in Chapter 6. The R_F values are measured and any slight adjustment in activity that may be necessary is achieved by changing the relative humidity of the storage chamber in which the plates are stored.

In column chromatography the same test dye mixture is eluted with benzene and the retention volumes are measured. Any slight adjustments are achieved by adding a small amount of water or dry (activated) adsorbent and allowing to equilibrate a further 24 hr. The use of naphthalene, eluted by n-pentane, is now gaining popularity as a reference standard since this system is more amenable to the wide range of sensitive detectors now coming into use.

Deviations from Equation (3.1) have been known to occur, for example, with solvents that are sufficiently polar to displace water from the adsorbent surface, with Florisil containing less than 1% water[2], and in cases where extremely-fine-pore adsorbents have been used with solvents of large molecular diameter. Such a case is solute elution from fine-pore silica (average pore diameter 20 Å) with carbon tetrachloride.

3.4.4. Choice of Mobile Phase

Having chosen the adsorbent type and standardized its activity commensurate with a maximum linear capacity, the most important method of adjusting $K°$ to give an optimum value of R_F or V_r is to choose a mobile phase with the correct eluent strength. The effect of the mobile phase upon $K°$ is given by the expression $(S° - A_s\varepsilon°)$ in Eq. (3.1), where $S°$ is the sample adsorption energy on an adsorbent of standard activity ($\alpha = 1$), A_s is proportional to the molecular area of the solute molecule, and $\varepsilon°$ is defined as the eluent strength. The greater the eluent strength, the smaler is the value fo $K°$ for a given solute and adsorbent.

Values of $\varepsilon°$ (relative to n-pentane, for which $\varepsilon°$ is defined as zero) for numerous solvents have been determined on alumina[2]. Placing the solvents

Table 3.4. Eluotropic Series for Polar Adsorbents [a]

Solvent	$\varepsilon_o(Al_2O_3)$	Viscosity, cP, 20°	RI	UV cutoff, nm	Boiling point,°C
Fluoroalkanes	−0.25	—	1.25	—	—
n-Pentane	0.00	0.23	1.358	210	36
i-Octane	0.01	0.54	1.404	210	118
n-Heptane	0.01	0.41	1.388	210	98.4
n-Decane	0.04	0.92	1.412	210	174
Cyclohexane	0.04	1.00	1.427	210	81
Cyclopentane	0.05	0.47	1.406	210	49.3
Carbon disulfide	0.15	0.37	1.626	380	45
Carbon tetrachloride	0.18	0 97	1.466	265	76.7
Amyl chloride	0.26	0.43	1.413	225	108.2
i-Propyl ether	0.28	0.37	1.368	220	69
i-Propyl chloride	0.29	0.33	1.378	225	34.8
Toluene	0.29	0.59	1.496	285	110.6
n-Propyl chloride	0.30	0.35	1.389	225	46.6
Chlorobenzene	0.30	0.80	1.525	280	132
Benzene	0.32	0.65	1.501	280	80.1
Ethyl bromide	0.37	0.41	1.424	225	38.4
Ethyl ether	0.38	0.23	1.353	220	34.6
Chloroform	0.40	0.57	1.443	245	61.2
Methylene chloride	0.42	0.44	1.424	245	41
Tetrahydrofurane	0.45	0.55	1.408	220	65
Ethylene dichloride	0.49	0.79	1.445	230	84
Methylethylketone	0.51	0.43	1.381	330	79.6
Acetone	0.56	0.32	1.359	330	56.2
Dioxane	0.56	1.54	1.422	220	104
Ethyl acetate	0.58	0.45	1.370	260	77.1
Methyl acetate	0.60	0.37	1.362	260	57
Amyl alcohol	0.61	4.1	1.410	210	137.3
Dimethyl sulfoxide	0.62	2.24	1.478	270	190
Aniline	0.62	4.4	1.586	325	184
Nitromethane	0.64	0.67	1.394	380	100.8
Acetonitrile	0.65	0.37	1.344	210	80.1
Pyridine	0.71	0.94	1.510	305	115.5
i-Propanol	0.82	2.3	1.38	210	82.4
Ethanol	0.88	1.20	1.361	210	78.5
Methanol	0.95	0.60	1.329	210	65.0
Ethylene glycol	1.11	19.9	1.427	210	198
Acetic acid	Large	1.26	1.372	251	118.5

[a] Data taken partly from Ref. 2.

in order of increasing eluent strength gives rise to the so-called eluotropic series (Table 3.4). The order is similar for all polar adsorbents, but is reversed for graphitized carbon (Table 3.5) because nonpolar molecules are preferentially adsorbed on this adsorbent.

Table 3.5. Eluotropic Series for Graphitized Carbon [a]

Water
Methanol
Ethanol
Acetone
Propanol
Ethyl ether
Butanol
Ethyl acetate
n-Hexane
Benzene

[a] Eluent strength increases downward.

In practice, the best way of choosing the correct eluent is by carrying out trial TLC on microscope slides. A pair of microscope slides, back-to-back, is dipped into a slurry of adsorbent (e.g., silica gel suspended in chloroform–methanol mixtures). The slides are separated and allowed to dry quickly. Then, 2 μl of a 1 % solution of the sample to be separated is applied to the plate and developed in a solvent of low eluent strength. This is usually achieved in 2–3 min. The plates are then allowed to dry and the chromatogram revealed by placing the plate in an iodine chamber. The experiment is repeated using solvents of increasing strength, using binary mixtures if necessary, until the R_F values of the components of interest lie between 0.3 and 0.8.

In addition to choosing a mobile phase of the correct eluent strength, the following factors have to be taken into consideration: (a) volatility, (b) method of detection of solute, (c) viscosity, (d) solvent demixing, (e) solubility of solute, and (f) effect on adsorbent.

In TLC the mobile phase must be sufficiently volatile so that it can be removed from the adsorbent to allow detection of the separated solutes to take place. This is particularly important when the mobile phase reacts with the revealing agent; e.g., most organic substances react with the dichromate–sulfuric acid reagent. With high-boiling polar mobile phases prolonged heating of the plate in an oven may be necessary before the last traces of eluent are removed. The mobile phase must also be free from nonvolatile impurities in cases where recovery of the separated solute is required. The removal of large amounts of high-boiling solvent can lead to substantial loss or decomposition of the solute.

In column chromatography the choice of mobile phase must be compatible with the choice of detector. Transport ionization detectors, for example, require separation of mobile phase from the solute by evaporation before the latter is carried forward into the detector. Where an ultraviolet spectrophotometer is used as a detector, the mobile phase must be transparent over the wavelength range where the solute is to be measured. Solvents that are particularly useful are saturated hydrocarbons, halogenated hydrocarbons, ethers, acetonitrile, and alcohols. Ethers must be free from peroxides and oxidation inhibitors. They should therefore be distilled immediately before use, utilizing only the first 75 vol % collected. With differential refractometers maximum sensitivity will occur when a solvent is chosen such that the difference in refractive index between solvent and solute is maximized.

As discussed in an earlier chapter, the viscosity of the solvent must be as low as possible to achieve acceptable efficiencies in the shortest times. Solvents which are particularly suitable from this point of view include n-pentane, carbon disulfide, 1- and 2-chloropropane, diethyl ether, acetone, acetonitrile, and methanol.

Solvent demixing often occurs when using binary or more complex mixtures and is readily observable in TLC. It occurs when the strongest solvent is preferentially adsorbed, leaving the advancing solvent front richer in the least polar component. Such demixing is generally undesirable since two or more components may travel unresolved with the secondary solvent front. It is more likely to occur when two solvents of widely different polarity are mixed, e.g., chloroform–methanol. If binary mixtures must be used, therefore, it is better to mix two solvents of similar polarity. As an example, 5% diethyl ether in heptane is an excellent mobile phase for the separation of polynuclear aromatics on alumina. On the other hand, 1% ethanol in heptane, while of similar eluent strength, is not a good mobile phase for this separation, since demixing occurs.

The solvent must be able to dissolve the solute at the working temperature of the separation. Even in difficult cases this can usually be achieved by selecting suitable solvent mixtures of the correct eluent strength. For example, a suitable solvent system for the chromatography of a substance containing a long hydrocarbon chain attached to a strongly polar group may require a solvent system containing a hydrocarbon such as toluene to provide the solvent power and ammonia to provide the correct eluent strength, together with a "bridging" solvent such as isopropanol to provide a homogeneous mobile phase.

When using deactivated adsorbents care must be taken that the mobile phase does not reactivate the adsorbent. Water-miscible solvents such as alcohols will strip the adsorbent of surface-adsorbed water, while even solvents as weak as n-pentane will gradually reactivate magnesia, silica, and alumina. To prevent this, a small amount, usually 0.1% or less, of water must be added to the eluent to maintain the adsorbent's activity. Regular standardization of the column as described in the previous section will determine whether the water content of the eluent should be decreased or increased. With water-miscible solvents addition of the requisite amount of water is all that is required. With water-immiscible solvents such as pentane shaking with water is unsatisfactory because dissolution of water is very slow and water droplets form in the solvent. A satisfactory method is to percolate it through a column (100 × 4 cm is satisfactory) packed with silica gel impregnated with about 30 % by weight with water.

3.5. VARIATION OF ADSORPTION COEFFICIENT $K°$ WITH SOLUTE STRUCTURE

The most important parameter affecting the value of $K°$ is the molecular structure of the solute. The effect of molecular structure upon the value of $K°$ in Eq. (3.1) is described by $S°$, the solute adsorption energy. This is approximately equal to the sum of the adsorption energies $Q_i°$ of the individual functional groups within the molecule[2].

Group adsorption energies of a number of functional groups have been determined for different adsorbents, some values on alumina and silica being given in Table 3.6. The values for Florisil are usually identical to those of silica. The higher the value of the group adsorption energy, the more strongly is the molecule adsorbed.

A consequence of the low value for the methylene group is that members of a homologous series have nearly identical adsorption energies. In adsorption chromatography, therefore, in contrast to partition chromatography, combounds are separated by type, but not by molecular weight.

The values of group adsorption energies may be modified by intramolecular electronic effects, steric effects (e.g., planar molecules are more easily adsorbed than nonplanar ones, *trans* isomers more than *cis*; *ortho*-substituted aromatic rings less strongly than *meta* or *para* ones), and chemical interaction between adjacent functional groups (e.g., hydrogen bonding reduces adsorption energies).

For a detailed discussion of the effect of sample structure on $K°$ values the reader is referred to Ref. 2.

3.6. INDIVIDUAL ADSORBENTS

3.6.1. Silica Gel

Adsorbents with the general formula $SiO_2–xH_2O$ have been described as silica, silica gel, or silicic acid. Silica gel is probably the most commonly used chromatographic adsorbent at the present time. This is a consequence of its high sample capacity, inertness to most labile solutes, and commercial availability. There is, therefore, a large amount of literature on the applications of silica gel as a chromatographic adsorbent. Unfortunately, because

Table 3.6. Adsorption Energies $Q°$, of Various Sample Groups[a]

Group	Alumina			Silica			Florisil[b]		
	X,Y =Ar	X=Al Y=Ar	X,Y =Al	X,Y =Ar	X=Al Y=Ar	X,Y =Al	X,Y =Ar	X=Al Y=Ar	X,Y =Al
X—CH₃ methyl	0.06	—	-0.03	0.11	—	0.07	0.10	—	-0.01
X—CH₂—Y methylene	0.12	0.07	0.02	0.07	0.01	-0.05	0.19	0.10	0.01
X—Cl chloro	0.20	—	1.82	-0.20	—	1.32	-0.20	—	1.74
X—F fluoro	0.11	—	1.64	-0.15	—	1.30	-0.15	—	1.54
X-—Br bromo	0.33	—	2.00	-0.17	—	1.32	-0.17	—	1.94
X—I iodo	0.51	—	2.00	-0.15	—	1.28	-0.15	—	1.94
X—SH mercapto	8.70	—	2.80	0.67	—	1.70	0.67	—	1.70
X—S—S—Y disulfide	?	~1.1	2.70	?	0.94	1.90	?	0.94	1.90
X—S—Y sulfide	0.76	1.32	2.65	0.48	1.29	2.94	?	1.30	2.94
X—O—Y ether	1.04	1.77	3.50	0.87	1.83	3.61	0.87	1.81	3.61
X—N—Y tertiary amine	?	2.48	4.40	?	2.52	~5.8	?	?	?
X—CHO aldehyde	3.35	—	4.73	3.48	—	4.97	3.35	—	4.97
X—NO₂ nitro	2.75	—	5.40	2.77	—	5.71	3.07	—	5.71
X—C≡nitrile	3.25	—	5.00	3.33	—	5.27	3.33	—	5.27
X—CO₂—Y ester	4.02	3.40	5.00	4.18	3.45	5.27	4.18	3.45	5.27
X—CO—Y keto	4.36	3.74	5.00	4.56	4.69	5.27	4.56	4.32	5.27
X—OH hydroxyl	7.40	—	6.50	4.20	—	5.60	4.20	—	5.60
X—C═N—Y imine	4.14	4.46	6.00	?	?	?	?	?	?
X—NH₂ amino	4.41	—	6.24	5.10	—	8.00	?	?	?
X—SO—Y sulfoxide	?	4.0	6.70	?	4.2	7.2	?	4.2	7.2
X—COOH carboxylic acid	19	—	21	6.1	—	7.6	6.1	—	7.6
X—CONH₂ amide	6.2	—	8.9	6.6	—	9.6	6.6	—	9.6
C═aromatic carbon	0.31	0.31	0.31	0.25	0.25	0.25	0.18	0.18	0.18

[a] Reproduced from Ref. 8, by courtesy of Elsevier Publishing Co. and L. R. Snyder.
[b] Assumes $\alpha = 1.00$ for 1% H_2O-Florisil.

of the many commercial sources, wide variations in properties exist, e.g., surface pH, pore-size distribution, and surface area.

Silicas are usually prepared either by acid precipitation from metal silicate solutions, especially sodium silicate, or by the hydrolysis of silicon compounds, such as silicon tetrachloride, in the liquid or vapor phase. The pore size, surface area, and nature of the surface vary according to the method of preparation. Variation of the solution pH during the acid gelation of sodium silicate, for example, produces silicas with surface area varying from about 200 m^2 g^{-1} (pH about 10) to 800 m^2 g^{-1} ($pH \leq 4$). Most chromatographic silicas, especially those used for TLC, have surface areas of 300–600 m^2 g^{-1} and pore diameters of 100–250 Å, and are classed as large-pore silicas. They have a semicrystalline structure and a relatively uniform surface covered predominantly with free hydroxyl groups (4–5 hydroxyls per 100 Å^2 of surface). A number of small-pore silicas are now commercially available whose average pore diameters are less than 100 Å and whose surface areas are in excess of 500 m^2 g^{-1}. They have an irregular, amorphous structure, their surfaces being covered predominantly with reactive and bound hydroxyl groups (see below). Silicas are also available in a series of controlled pore sizes covering the range 100–2500 Å and are useful for separating polymers by the molecular exclusion principle. These are further discussed in Chapter 5.

As has already been indicated, a silica surface that has not been heated for long periods in excess of 400 °C is covered, to a greater or less extent, by hydroxyl groups. It is the presence of these surface hydroxyl groups that is responsible for the selective adsorption properties of silica. Thus silica adsorbs unsaturated, aromatic, or polar molecules by hydrogen bonding, the solute functioning as an electron donor. Carbon–carbon double bonds contribute somewhat less to sample adsorption energy on silica compared with other polar adsorbents. Aromatic hydrocarbons and compounds differing only in their relative degree of unsaturation are better separated on other polar adsorbents such as alumina.

The silica surface is weakly acidic (pH 3–5) and therefore there is a tendency toward preferential adsorption of strongly basic compounds ($pK_B < 5$) relative to their adsorption on neutral or basic adsorbents. Sometimes the silica surface is found to be strongly acidic. This is due to contamination by acids left over from the gelation step. Such silicas should be cleaned by repeated washing with distilled water; otherwise chemisorption of bases or reaction of acid-sensitive solutes will occur.

The nature of the silica surface has been studied and reviewed by a number of workers (e.g., see Ref. 35). Conflicting views on the relationships between silica processing, surface structure, and chromatographic properties still exist, but many discrepancies now seem to have been resolved, mainly due to the efforts of Snyder and his coworkers (see Ref. 2). A greatly simplified summary of their conclusions is given below.

An air-dried silica surface that has not undergone previous heat treatment contains physically adsorbed water. Heating at temperatures between 150 and 200 °C drives off most of the adsorbed water leaving a surface containing hydroxyl groups. The hydroxyl groups can be divided into three types as shown in Fig. 3.6. The surface of large-pore silicas consists predominantly of free hydroxyls (Fig. 3.6a), while that of small-pore silicas contains reactive (Fig. 3.6c) and bound (Fig. 3.6b, c) hydroxyls. The activity of the different types of site increases in the order: bound < free < reactive hydroxyls. This means that the surface activity of small-pore silicas is greater than that of wide-pore silicas because of the greater concentration of reactive hydroxyls in the small-pore silica. However, the effect of adding water to an activated silica surface is to deactivate the reactive hydroxyls of a small-pore silica first, leaving a surface of bound hydroxyls. Corresponding deactivation of a large-pore silica leaves a surface of free hydroxyls. Consequently, the surface activity of a heavily deactivated small-pore silica is less than a similarly deactivated large-pore silica. The effect of water deactivation on the linear capacity of both large- and small-pore silicas has already been discussed (Fig. 3.4).

To classify the surface structure of a silica gel we need to know the ratio of reactive to total hydroxyls (S_r/S_t) and the surface area. A method for determining these has been proposed by Snyder and Ward (see Ref. 2). Reactive hydroxyls are determined by selective silanization with trimethylchlorosilane, and total hydroxyls are determined by complete silanization

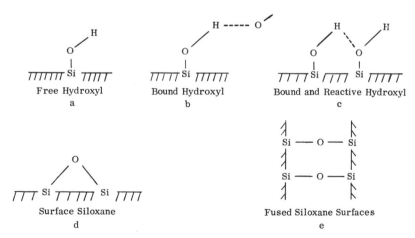

Fig. 3.6. The nature of the silica surface. [Reproduced by courtesy of Marcel Dekker, Inc., and L. R. Snyder[2].]

Table 3.7. Differences Between Small- and Large-Pore Silicas [a]

Property	Small-pore silicas	Large-pore silicas
Average pore diameter	< 100 Å	> 150 Å
Specific surface area	> 500 m² g⁻¹	< 600 m² g⁻¹
Surface structure	Irregular, amorphous, heterogeneous	Semicrystalline, uniform
Predominant hydroxyl types	Reactive plus bound	Free
Surface activity α		
Activated silica	> 1.00	< 0.9
Deactivated silica[b]	~ 0.6	~ 0.7
Variation in α with increase in silica activation temperature (150–400°)	Decreases	Little change
Linear capacity		
Activated silica	$\sim 10^{-4} \times$ g g⁻¹	$\sim 2 \times 10^{-4}$ g g⁻¹
Deactivated silica	25×10^{-4} g g⁻¹	4×10^{-4} g g⁻¹

[a] Reproduced in part from (²) by courtesy of Marcel Dekker, Inc., New York and L.R. Snyder.
[b] 60% Coverage of surface by water (2.1% water per 100 m² g⁻¹ of surface).

with hexamethyldisilazane. The silica surface area can be calculated from an experimental value of S_t. The differences in properties between small- and large-pore silicas are summarized in Table 3.7.

If activated silicas are heated at temperatures between 200 and 400°C reactive hydroxyl groups condense to liberate water to form surface siloxane groups (Fig. 3.6d). Similarly, free hydroxyl groups begin to migrate about the surface, form transient reactive groups, and then decompose to form surface siloxane groups. The rehydration of such siloxane groups is difficult and requires prolonged heating at 95°C with water. Heating silica surfaces at temperatures above 400°C causes condensation between adjacent planes (Fig. 3.6e), resulting in decrease in surface area and loss of selectivity.

The conclusions to be drawn from this account are that the chromatographic properties of silica can be expected to vary widely according to the method of manufacture and subsequent thermal treatment. To obtain reproducible results, it is therefore necessary to have a knowledge of the pore diameter and surface properties, first, to ensure that adsorbents with similar selectivities are being used, and second, to give a guide to the further treatment required to produce a standardized adsorbent of optimized linear capacity.

3.6.2. Alumina

After silica, alumina is the most popular adsorbent in use today and is readily available commercially. It is a polar adsorbent like silica, and the order of elution of solutes on the two adsorbents is generally similar. There are three important practical differences from silica:

(a) The adsorption energy of a solute molecule containing isolated and conjugated carbon–carbon double bonds is greater on alumina. Consequently, alumina is preferred for the separation of mixtures of aromatic hydrocarbons, because their adsorption coefficients cover a wider range.

(b) Alumina contains a number of strongly basic sites and therefore shows a preferential adsorption of acidic samples. Strong acids ($pK_a \leq 5$) are chemisorbed, but weaker acids can be separated in order of their pK_a values, especially when basic eluents are used.

(c) Activated alumina cannot be used for some separations because certain solute types undergo chemical reaction at the reactive sites[9]. Examples are: salt formation with acids, saponification of esters and anhydrides, condensation reactions with aldehydes and ketones, elimination reactions with loss of hydrogen halides, isomerization and polymerization of olefins, oxidation reactions, and complex formation.

Alumina can exist in a number of crystalline forms, depending upon the method of preparation and thermal history. Most commercial aluminas prepared for chromatography probably consist mainly of the γ-form, but also contain small amounts of other crystalline forms, e.g., η, χ, ϱ. All these forms have similar chromatographic properties. On heating to 900–1000 °C these are converted to other crystalline modifications[1], (θ, δ, K). On heating to temperatures higher than 1100 °C all aluminas are converted to α-alumina, which is chromatographically inactive, presumably on account of its low surface area and different lattice structure. The ideal crystal of γ-alumina consists of layers of large oxide ions (O^{2-}) with small aluminum ions (Al^{3+}) occupying three out of every four holes between the oxide ions. At room temperature up to a monolayer of water is readily adsorbed onto the alumina surface, each water molecule being bound to two surface oxide ions. On heating a hydrated alumina to 300–400 °C most of the adsorbed water is driven off, with the remainder of the water reacting with the surface to form hydroxyl groups, up to six hydroxyl groups per 100 Å^2 of surface being formed. This is the form in which alumina for chromatography is generally used. The linear capacity is increased by the addition of water corresponding to 50 % coverage, although many workers have deactivated their adsorbents with greater amounts of water.

Activated alumina has been observed by electron microscopy[10] to consist of a system of regular, cylindrical micropores of 27 Å diameter arranged hexagonally, in addition to two types of random macropore, those within the particles of alumina that make up the granules, and those between such particles. The surface area of a typical chromatographic alumina lies in the range 100–200 m^2 g^{-1}.

Various proposals have been made regarding the nature of the active sites and the corresponding adsorption interactions on the alumina surface.

On heating a hydrated alumina to more than 400°C, hydroxyl groups are gradually removed, but surface hydroxyl groups are not completely eliminated even by heating under vacuum at 800–1000°C. Nevertheless, chromatographic activity increases with increasing temperature of activation up to 1100°C. This is taken as evidence that unlike the case for silica, surface hydroxyl groups do not play an important role in the adsorption of solutes onto alumina. Snyder[2] recognizes three distinct types of adsorption site:

(a) Acidic or electrophilic field sites interact with solutes possessing regions of high electron density. This is the most common adsorption mechanism encountered, showing that the alumina surface behaves as an acid toward most solute types.

(b) Basic or nucleophilic sites (probably oxide ions) are responsible for the preferential adsorption of acids relative to other adsorbents.

(c) Electron-acceptor (charge-transfer) sites form complexes with easily polarized aromatic molecules like naphthalene. The exact nature of these sites is not yet known.

3.6.3. Magnesium Silicates

These adsorbents are coprecipitates of silica and magnesia. A well-known commercial product is Florisil (Floridin Co., Pittsburgh, Pa.), which is a white material containing 84 % of silica. Reviews of its use and chromatographic properties are given in Refs. 11 and 12.

The product, as received, has an average pore diameter of 62 Å and a surface area of 300 m^2 g^{-1}. Activated Florisil, obtained by heating to 400° for 16 hr, possesses strongly acidic sites on its surface, and in addition to chemisorbing organic nitrogen bases, also partially but irreversibly adsorbs other compound types such as esters and aromatic hydrocarbons. Deactivation by the addition of up to 1 % water preferentially covers these acidic sites. Further deactivation produces an adsorbent whose chromatographic properties are intermediate between those of silica and alumina.

3.6.4. Magnesia

Magnesia (magnesium hydroxide or oxide) is a polar, basic adsorbent which has recently been evaluated by Snyder[13]. It is available as a very fine white powder suitable for both TLC and high-efficiency chromatography.

It is believed that surface hydroxyl groups play an important part in the adsorption mechanism. On heating to 150°C, varying amounts of physically adsorbed water are lost. On heating to 350°C the activity of magnesium hydroxide sharply increases due to loss of surface hydroxyls and formation of oxide groups. On further heating the activity of the magnesia surface is reduced and above 1000°C becomes completely inactive. Activated magnesia chemisorbs aromatic compounds. To avoid this, a satisfactory adsorbent may be prepared by activation at 500 °C for 16 hr followed by deactivation with 3–7% water. The linear capacity of 3% H$_2$O–MgO was found to be approximately 2×10^{-5} g g^{-1}[13].

Deactivated magnesia is readily reactivated by dry solvents such as pentane. Water-saturated solvents must therefore always be used if chemisorption is to be avoided. The selectivity of deactivated magnesia is similar to that of silica or alumina. However, compounds containing carbon–carbon unsaturation are much more strongly held on magnesia than on alumina. Magnesia is therefore a valuable adsorbent for the separation of compound classes differing only by degree of unsaturation, e.g., olefins from diolefins, polynuclear aromatics, etc. On account of its high surface pH, acids are chemisorbed.

3.6.5. Modified Adsorbents

The properties of silica gel or of other polar adsorbents may be modified by incorporating a complexing agent into the adsorbent. For example, the separation of olefinic from saturated hydrocarbons is much better if silica gel is first impregnated with silver nitrate solution. Further examples of modifiers are given in Table 3.8. In general, 1–10% solutions of the complexing agent in water or acetone are slurried with the adsorbent. The slurries are either directly spread over the plates in the usual manner of else dried at 110 °C in an oven before being packed into columns.

3.6.6. Porous Layer Beads

Rigid nonporous supports such as glass beads covered by a thin porous layer of adsorbent (and on which a liquid stationary phase can be deposited—

Table 3.8. Modified Adsorbents

Complexing agent	Selective for	References
0.1–0.5 N acids or bases	pH-sensitive compounds	(14)
Silver nitrate	Olefinic or acetylenic materials	(15-17)
Boric acid, sodium borate, sodium arsenite, basic lead acetate, sodium metavanadate, sodium molybdate	Polyhydroxyl compounds	(18)
Caffeine, 2,4,7-trinitrofluorene, picric acid, trinitrobenzene	Polynuclear aromatic hydrocarbons	(19-22)
Sodium bisulfite	Aldehydes	(23)
Ferric chloride	Oxines	(24)
Copper sulfate	Amines	(25)
Zinc ferrocyanide	Sulfonamides	(26)

see Chapter 4) are currently being developed for use in high-efficiency columns. A wider range than that shown in Table 3.2 will probably be commercially available by the time this book is published.

The materials are hard, free-flowing, and have regular dimensions so that they can easily be dry-packed to give columns with efficiencies of the order of 5000 plates per meter. The flatness of their HETP/mobile phase velocity curves (see Fig. 4.6) indicates that such packings can be used at high flow rates giving fast analyses with little sacrifice in column efficiency. This is made possible because they can withstand pressures up to at least 500 atm, thus enabling columns to be operated at high flow rates.

Horvath et al.([27,28]) coated glass beads with a thin skin (pellicule) of ion-exchange resin by copolymerizing styrene–divinyl benzene directly on the beads, followed by chemical conversion to the appropriate ionic form.

Kirkland([29,30]) introduced his "controlled-surface-porosity" supports, which consist of glass beads with a porous surface of controlled thickness and pore size (Zipax*). A support of this type with an average pore diameter of about 1000 A has a surface area of 0.65 m^2 g^{-1}. They are very weak adsorbents with a low capacity and are therefore more useful as support materials for partition chromatography (see Chapter 4). A packing with cation-exchange properties has also been prepared in which the porous layer consists of a fluoropolymer containing free sulfonic acid groups. The packing can be used at elevated temperatures with a variety of mobile phases and can withstand column inlet pressures of greater than 200 atm. The cation-exchange capacity was 3.5 μeq g^{-1}. Similarly, a strongly basic, controlled-surface-porosity anion-exchange packing has been prepared containing tetraalkylammonium groups. The capacity of this particular packing was found to be 12 μeq g^{-1}.

Corasil (Waters Associates) is an adsorbent consisting of glass beads coated with either a single or double layer of porous silica (Corasil I and II, respectively). They are activated by heating overnight at 110°C and then deactivated by adding up to 0.5% by weight of water. Heating above 300°C produces a permanent reduction of surface activity. The linear capacity of Corasil II (surface area 14 m^2 g^{-1}) is an order of magnitude less than that of conventional silica gel. A recent evaluation([31]) illustrates high-speed analyses of phenolic and amine antioxidants.

3.6.7. Carbon

Carbon is an excellent adsorbent but, on account of its ill-defined properties and its color, which makes visual detection of solute bands

*Registered trademark, E. I. du Pont de Nemours & Co., Inc.

difficult, it has so far found limited use in adsorption chromatography. It strongly adsorbs aromatics and high-molecular-weight compounds and it has therefore been used as a "cleaning-up" agent, e.g., for removing high-molcular-weight compounds from complex mixtures to enable low-molecular-weight materials to be more easily analyzed. An example is the removal of high-molecular-weight carbohydrate material from plant residues in the search for insecticides.

The color of carbon has almost completely precluded its use in TLC because of detection difficulties. However, with the types of column and detection systems described in this book the field is wide open for a reappraisal of the use of certain forms of activated carbon as adsorbent. The evaluation of previously qublished work using active carbon as an adsorbent is often difficult because of the ill-defined nature of the carbon used. However, the nature of the carbon surface is now more fully understood (for recent reviews see the list of further reading) and carbons with more precisely defined properties are now available. Examples are Graphon, a completely nonpolar charcoal with a hydrophobic surface, and the more polar Spheron 6. Both are available from Cabot Corporation. It is the nonpolar and hydrophobic nature of active carbons which distinguishes them from the metal oxides and makes them particularly attractive as adsorbents.

Carbon occurs naturally in two allotropic forms—diamond, which need not concern us any further, and graphite. Graphite is a well-defined crystalline form of elemental carbon consisting of layers 3.35 Å apart of carbon atoms joined by sp^2 covalent bonds in a fused hexagonal ring system. The layers are held together by the relatively weak van der Waals forces.

Adsorbent carbons can be divided into two main categories—charcoal and carbon black. Charcoals are made by the destructive distillation of organic matter such as wood or bone. Activation is achieved by slow oxidation at elevated temperatures with air, steam, carbon dioxide, or chlorine, or by impregnation by salts, acids, or alkalis followed by calcination. Charcoals so produced possess heterogeneous surfaces containing inorganic atoms in addition to organic functional groups, making adsorption data difficult, if not impossible, to interpret. The adsorbent properties of charcoals from different suppliers will therefore differ, and variations in the properties from a given supplier can also be expected from batch to batch.

Carbon blacks are formed by the incomplete combustion of hydrocarbons. They are microcrystalline materials consisting of graphite-like layers 3.6 Å apart stacked in packets of 3–30 layers about 10–100 Å thick. However, the layers often contain tetrahedrally bonded carbon atoms giving rise to cross-linking and causing lattice defects. The unsatisfied bonds at the edges of the graphitic layer planes are very reactive and during manufacture combine with foreign atoms or groups. For example, active carbons formed by low-temperature oxidation usually possess acidic oxide surfaces due to the presence of carboxyl, carbonyl, and phenolic groups. Basic groups have also been observed.

In general, the surface of carbon black is hydrophobic, nonpolar, and shows poor specificity toward functional groups. However, owing to the presence of inorganic substances and polar functional groups, the surface also contains hydrophilic sites. It is due to the presence of these sites that many carbon blacks possess adsorption properties resembling those of the metal oxides. Adsorption on such sites involves electrostatic forces and hydrogen bonding. Ion-exchange properties are also exhibited by such carbons, e.g., they adsorb certain organic salts with the liberation of acids or bases.

On heating a carbon black such as Spheron 6 to a temperature of 3000°C under inert conditions, hydrophilic functional groups are lost, while the graphitic layer planes undergo a reorientation to form a more crystalline material. The resulting surface has a more hydrophobic, nonpolar character and specificity toward functional groups is absent. In contrast to the metal oxides, the main contribution to the adsorption energy between solutes and this material (Graphon) is due to dispersion forces.

The differences in properties between Spheron 6 and Graphon have been demonstrated by a number of workers. On the more polar Spheron 6, methanol is adsorbed more strongly than benzene, while in the case of Graphon the reverse is true. On Spheron 6 adsorption is by hydrogen bonding between the polar group of the adsorbent and the hydroxyl group of the solute, the hydrocarbon chain of the alcohol extending vertically outward into the solution. On Graphon the alcohol lies flat along the surface of the adsorbent. It has also been observed that aliphatic acids are adsorbed on Graphon in the dimeric form and lie with their long axes parallel to the surface. Similar orientations have been observed with alkyl benzenes and methyl esters of monocarboxylic acids.

Many commercial carbons have properties that fall between the two extremes, i.e., both surface oxygenated groups and the graphite structure are present. The adsorption properties of the oxygenated carbons resemble those of the metal oxides, but, due to the graphite structure, aromatic compounds are adsorbed more strongly than the corresponding aliphatic derivatives. This gives rise to a different eluotropic series than that of the metal oxides, an example of which is given in Table 3.5. It will be noted that the solvent strength increases with increasing size of solvent molecule. Aromatic solvents will also be seen to be stronger than the corresponding aliphatic solvents.

Active carbons are spherical, porous particles, their surface areas and pore diameters depending upon the method of preparation and subsequent

Table 3.9

Pore type	Pore diameter, Å	Surface area, $m^2 g^{-1}$
Macropores	2000–1,000,000	0.5–2
Transitional pores	30–2000	80–400
Micropores	<30	>400

activation treatment. In general, three pore types can be recognized, macropores, transitional pores, and micropores, the characteristics of each being given in Table 3.9.

The low surface area of the macropores indicates that such pores play a negligible role in adsorption. However, they do provide access channels to the more important transitional pores and micropores.

The surface area and pore diameter of transitional pores compare with those of silica gel or alumina, and carbons of this type have been widely used for removing large molecules or colloidal material from solutions containing smaller molecules, e.g., in the cleaning up of biological residues. Molecules are adsorbed on the surface to form a monolayer.

The pore diameter of microporous carbons is of the same order as the size of many organic molecules. Adsorption energies are relatively high, giving rise to slow mass-transfer rates. This in turn can lead to poor separations.

3.6.8. Ion Exchangers

Ion exchangers are naturally occurring or synthetic, insoluble polyelectrolytes having a porous structure which can take up positive or negative ions from an electrolyte solution in contact with them in exchange for an equivalent amount of its own ions which are liberated into solution. Extensive use has been made of this property for the removal, concentration, exchange, or analysis of inorganic ions in aqueous solutions. These aspects, which are beyond the scope of this book, are adequately discussed elsewhere (see the list of further reading). Of more concern to us is the property of ion exchangers to act as acidic or basic adsorbents in nonionic organic media[32].

The most common type of ion exchanger in use today consists of an irregular, macromolecular, three-dimensional network of hydrocarbon chains bearing ionizable groups. The matrix is a copolymer of styrene and divinyl benzene into which various functional groups are introduced either before or after polymerization to give the resin its ion-exchange properties. The products are classified according to the type of active group as follows:

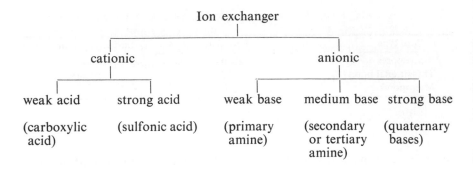

The so-called "macroreticular" ion-exchange resins have a rigid, macro-porous structure with pore diameters up to 800 Å superimposed upon the normal gel structure. Their physical structure is therefore similar to silica or alumina. Otherwise, they are chemically similar to the conventional polystyrene–divinyl benzene based resins generally used in aqueous solution. As a consequence of their macroporous structure they can be used in solvents in which resins do not swell (i.e., nonpolar organic solvents) and for the adsorption of large organic molecules or ions.

A number of macroreticular resins are now commercially available. Their properties (Table 3.10) can be compared with a commonly used conventional resin. As a typical example[33], Amberlyst A-15, a sulfonic acid cation exchanger, has a specific surface area of 42.5 m^2 g^{-1} compared with under 0.1 m^2 g^{-1} for the conventional resin Amberlite IR-120. Examination by electron microscopy revealed no internal pore structure for the conventional resin, but a definite porous structure for A-15, pores of 400–800 Å diameter being evident. Ion-exchange equilibria and kinetic studies showed

Table 3.10. Some Properties of Macroreticular Resins

	Amberlyst resins[a]					Amberlite[a]
Designation	A-15	XN1005	A-29	A-27	A-21	IR-120
Type	Cationic	Cationic	Anionic	Anionic	Anionic	Cationic
Ionic form	Hydrogen	Hydrogen	Chloride	Chloride	Free base	Hydrogen
Active group	–SO_3H	–SO_3H	Quaternary ammonium	Quaternary ammonium	Tertiary amine	–SO_3H
Capacity, meq g^{-1}	4.9	3.5	2.7	2.6	4.8	4.5
Capacity, meq ml^{-1}	2.9	—	1.0	0.7	1.6	1.7
Pore diameter, Å	200–600	—	200–600	400–800	700–1200	<5
Surface area, m^2 g^{-1}	40–50	122	40–50	60–70	20–30	<0.1

[a] Manufactured by Rohm and Haas Co., Philadelphia, Pa.

the macroreticular resin to be much less sensitive to the nature of the solvent than the conventional resin. They are able to withstand alternate wetting and drying without degrading in particle size, showing their superior physical stability over conventional resins. Furthermore, they show no marked deterioration when subjected to repeated changes from an aqueous to a non-aqueous environment.

Ion exchangers behave as acidic or basic adsorbents in organic media. Both physical and chemical adsorption are known to occur. For example, the acetate form of the anion exchanger Amberlyst A-29 physically adsorbs polar compounds such as pyrroles and phenols from hydrocarbon solutions. These solutes can be eluted from the resin with polar solvents such as pyridine or methanol. Acids, on the other hand, are chemisorbed on this resin and acidic solvents are required to elute them. In a similar fashion, the strong acid cation-exchange resin Amberlyst A-15 chemisorbs nitrogen bases from hydrocarbon solutions; they can be desorbed with basic solvents.

The selectivity of the resin can be made highly specific by making use of a form that will form a complex (ligand) with the solute[36]. For example, cation exchangers in either the Ag^+, Cu^{2+}, or Ni^{2+} forms have been used for the separation of amines and carboxylic acids; the Ag^+ form has also been used for isolating compounds with olefinic double bonds. Elution development is carried out with an agent which complexes less strongly than the substances to be separated; displacement development is carried out with an agent which complexes more strongly.

REFERENCES

1. Giles, C. H., MacEwan, T. H., Nakhwa, S. N., and Smith, D., *J. Chem. Soc.* 3973 (1960).
2. Snyder, L. R., *Principles of Adsorption Chromatography,* Dekker, New York (1968).
3. Snyder, L. R., *J. Chromatog.* 5, 430 (1961).
4. Snyder, L. R., *Separation Sci.* 1, 191 (1966).
5. Hesse, G., and Roscher, G., *Z. Anal. Chem.* 200, 3 (1964).
6. Hagdahl, L., and Holman, R. T., *J. Am. Chem. Soc.* 72, 701 (1950).
7. Snyder, L. R., *Advan. Anal. Chem. Inst.* 3, 251 (1964).
8. Snyder, L. R., *J. Chromatog.* 23, 388 (1966).
9. Hesse, G., *Z. Anal. Chem.* 211, 5 (1965).
10. Bowen, J. N., Bowrey, R., and Malin, A. S., *J. Catalysis* 1, 209 (1967).
11. Snyder, L. R., *J. Chromatog.* 12, 488 (1963).
12. *Floridin Technical Data and Product Specifications,* Floridin Co., Tallahassee, Fla.
13. Snyder, L. R., *J. Chromatog.* 28, 300 (1967).
14. Stahl, E., and Dumont, E., *J. Chromatog. Sci.* 7, 517 (1969)
15. Urbach, G., *J. Chromatog.* 12, 196 (1963).
16. Gupta, A. S., and Dev. S., *J. Chromatog.* 12, 189 (1963).
17. Chapman L. R., and Kuemmel, D. F., *Anal. Chem.* 37, 1598 (1965).
18. Morris, L. J., *J. Chromatog.* 12, 321 (1963).
19. Berg, A., and Lam, J., *J. Chromatog.* 16, 157 (1964).
20. Klemm, L. H., Reed, D., and Lind, C. D., *J. Org. Chem.* 22, 739 (1957).
21. Harvey, R. G., and Halonen, M., *J. Chromatog.* 25, 294 (1966).
22. Kessler, H., and Müller, E., *J. Chromatog.* 24, 469 (1966).

23. Adachi, S., *J. Chromatog.* **17**, 295 (1965).
24. Cawthorne, M. A., *J. Chromatog.* **25**, 164 (1966).
25. Brockmann, H., *Disc. Faraday Soc.*, **7**, 58 (1949).
26. Fogg, A. G., and Wood, R., *J. Chromatog.* **20**, 613 (1965).
27. Horvath, C. G., and Lipsky, S. R., *J. Chromatog. Sci.* **7**, 109 (1969).
28. Horvath, C. G., Preiss, B. A., and Lipsky, S. R., *Anal Chim.* **39**, 1422 (1967).
29. Kirkland, J. J., *Anal. Chem.* **41**, 218 (1969).
30. Kirkland, J. J., *J. Chromatog. Sci.* **7**, 361 (1969).
31. Majors, R. E., *J. Chromatog. Sci.* **8**, 338 (1970).
32. Webster, P. V., Wilson, J. N., and Franks, M. C., *Anal. Chim. Acta* **38**, 193 (1967).
33. Kunin, R., Meitzner, E. F., Oline, J. A., Fisher, S. A., and Frisch, N., *Ind. Eng. Chem. Prod. Res. Dev.* **1**, 140 (1962).
34. Heftmann, E., *Chromatography,* Reinhold, New York (1967).
35. Hockey, J. E., *Chem. Ind. (London)* 57 (1965).
36. Helfferich, F., Ion-exchange chromatography, *Advan. Chromatog.* **1**, 3 (1966).

FURTHER READING

Giles, C. H., and Easton, I. A., Adsorption chromatography, *Advan. Chromatog.* **3**, 70 (1966). (A discussion of isotherm shape and the nature of adsorption forces on silica, alumina, and carbon.)

Young, D. M., and Crowell, A. D., *Physical Adsorption of Gases*, Butterworths, London (1962). (A comprehensive review which gives an insight into the nature of physical adsorption.)

Snyder, L. R., *Principles of Adsorption Chromatography,* Dekker, New York (1968). (A comprehensive account of separation processes and basic principles of liquid–solid adsorption chromatography together with a systematic generalization of available data on the separation of nonionic organic compounds.)

Hayward, D. O., and Trapnell, B. M. W., *Chemisorption,* Butterworths, London (1964). (A comprehensive review which gives an insight into the nature of chemisorption.)

Snyder, L. R., R_F values in thin-layer chromatography on alumina and silica, *Advan. Chromatog.* **4**, 3 (1967). (A general theory for the correlation and prediction of R_F values in TLC. The theory is tested and confirmed by applying it to literature data.)

Everett, D. H., and Stone, F. S. (eds.), *The Structure and Properties of Porous Materials*, Butterworths, London (1958). (Contains chapters on the surface characteristics of carbons and silicas.)

Hockey, J. A., The surface properties of silica powders, *Chem. Ind. (London)* **1965**, 57;

Mitchell, S. A., surface properties of amorphous silicas, *Chem. Ind. (London)* **1966**, 924. (Accounts of the effects of different methods of preparation and subsequent thermal or chemical treatment on the nature and properties of the silica surface.)

Heftmann, E., *Chromatography,* Reinhold, New York (1967). (A general review of chromatography.)

Coughlin, R. W., Carbon as adsorbent and catalyst, *Ind. Eng. Chem. Prod. Res. Dev.* **8**, 12 (1969). (The structure and surface chemistry of carbon is related to its catalytic and adsorptive behavior.)

Dubinin, M. M., *Porous structure and adsorption properties of active carbons,* in *Chemistry and Physics of Carbon*, Walker, P. L., ed., Dekker, New York (1966), vol. 2, p. 51. (Basic studies.)

Zettlemoyer, A. C., and Narayan, K. S., Adsorption from solution by graphite surfaces, *Ibid.,* p. 197. (Some basic studies including a comparison of Graphon and Spheron 6.)

Amphlett, C. B., *Inorganic Ion Exchangers,* Elsevier, London (1964). (Covers the recently developed inorganic ion exchangers useful for high-temperature work. Mainly inorganic separations, though possibly applicable to nonaqueous work.)

Helfferich, F., *Inorganic Ion Exchangers,* McGraw-Hill, New York (1962). (A comprehensive account covering basic theory of ion exchangers.)

Inczédy, J., *Analytical Applications of Ion Exchangers,* Pergammon Press, Oxford (1966); Samuelson, C., *Ion Exchange Separations in Analytical Chemistry,* Wiley, New York (1963). (Useful practical books covering elementary theory and basic techniques of ion-exchange chromatography, including separations of organic mixtures.)

Helfferich, F., Ion exchange chromatography, *Advan. Chromatog.* **1,** 3 (1966). (A review of advances in ion-exchange chromatography during 1960–1965. Includes examples of organic separations in nonaqueous media and of ligand-exchange chromatography.)

Chapter 4

Liquid Partition Chromatography— Mechanism and Materials

4.1. INTRODUCTION

For the purpose of this book we will define partition processes in a very practical way. Partition of an organic solute occurs between a liquid mobile phase and an organic liquid absorbed in, or chemically bonded onto, a porous support. We will thus avoid complications which arise in trying to classify systems involving, for example water-modified adsorbents; are they partition or adsorption? For our purposes they are liquid–solid adsorption chromatographic processes.

In considering thermodynamic and practical aspects of liquid partition chromatography we need to start from a simple model system in which two liquid phases are in contact. An obvious example of such a system would be a separating funnel containing two immiscible liquids such as water and heptane. If a small quantity of a mixture of benzene and ethanol were introduced at the interface, most of the benzene would pass into the heptane bulk phase and most of the ethanol into the aqueous phase. A number of important points arise from the simple picture. First of all, what do we understand by immiscible liquids? Heptane and water are mutually soluble, albeit to a very limited extent (solubility of heptane in water is about 5×10^{-5} g cc^{-1} at 15 °C). Thus at equilibrium we are concerned with the distribution of solute between a dilute solution of heptane in water and a dilute solution of water in heptane, the concentration of each solution varying with temperature. We can see therefore that liquid partition chromatography will take us into the realms of the phase rule in considering the properties (and also the design of) suitable "immiscible" liquid phases.

The second point that emerges from our simple model relates to the means of contacting the two phases. In the example cited the solutes were introduced at the interface, and their rate of distribution would be controlled by rates of diffusion from the interface into the bulk of the contacting liquids. The realization of the equilibrium distribution would be enormously speeded up by vigorous shaking of the separating funnel, which would have the effect of increasing the contacting surface area as the phases become finely dispersed in each other. This, in fact, becomes a simple batch solvent extraction, a common laboratory practice. An example based on the use of a succession of batch solvent extractions in cascade to achieve eventually complete separation of two components is often used to illustrate the principle of countercurrent extraction. Countercurrent extraction is used industrially for the separation, purification, and production of organic materials. In the laboratory rather elaborate countercurrent extraction equipment is available, and has many applications, particularly in the field of biochemistry. Essentially, countercurrent separations represent a cumbersome, inefficient form of partition chromatography. The latter is a process in which intimate phase contacting is achieved by distributing the stationary liquid as a thin film over the surface of an inert, very-high-surface area, porous solid.

This simplified introduction to liquid partition has brought out several important matters of practical consequence which we will explore further.

In this chapter we discuss the thermodynamics of partition and their relationship to chromatographic behavior and measurements, and the phase rule and its bearing on the choice of phase pairs, and then move on to the very practical matter of the selection of suitable materials for partition chromatography.

4.2. THERMODYNAMICS OF LIQUID–LIQUID PARTITION

4.2.1. Relation between Partition Coefficient and Solution Properties

The art and science of partition chromatography rest on the ability to choose phase pairs of suitable selectivity that separation of components of interest is achieved. At present there is a large amount of art involved because the understanding of solution properties is not fully developed and knowledge of partition coefficients is essentially empirical. Nevertheless the ability to predict solution properties is being developed and is to be encouraged since it affords the only route to a full understanding and a maximal utilization of liquid partition processes.

The development of a very simplified theory follows from the following equations:

$$a_S/a_M = 1 \tag{4.1}$$

$$K = C_S/C_M \tag{4.2}$$

The former states the thermodynamic fact that at equilibrium the activity of a solute in two contacting, immiscible phases (a_S, a_M) is the same. The latter equation expresses that the partition coefficient of the same solute in the same two-phase system is given by the ratio of its concentrations in the two phases. Now, $a_S = C_S\gamma_S$ and $a_M = C_M\gamma_M$, where γ_S and γ_M are activity coefficients of solute in the stationary and mobile phases, respectively. Thus

$$C_S\gamma_S/C_M\gamma_M = 1 = K\gamma_S/\gamma_M, \quad \text{i.e.,} \quad K = \gamma_M/\gamma_S \tag{4.3}$$

In other words, the partition coefficient, and therefore the retention volume, is very simply related to the activity coefficients of the solute in the two phases. In a few cases activity coefficients are available in the literature and retention volumes are thus calculable. In some instances one activity coefficient can reasonably be assumed to be unity (e.g., a dilute solution of one alkane in another) and the other can therefore be calculated from a measured retention volume.

At this stage in the development of liquid partition chromatography, for practical purposes the situation is that one can sometimes *determine* activity coefficients (which are of value in chemical engineering calculations, for example) but their use in predicting separability of mixtures is rudimentary.

In the next sections, therefore, we inevitably turn to largely empirical approaches to the selection of phase pairs for LLC.

4.3. SELECTION OF PHASE PAIRS

We will consider the design of an LLC system from three standpoints: (i) its ability to achieve a desired separation (we will shortly refer to this as "selectivity"; (ii) its stability, in practical use; and (iii) its suitability from a dynamic standpoint in providing maximized resolving power and convenience of use.

4.3.1. Selectivity

The ability to separate molecules by a partition process arises from differences in their size, shape, or polar characteristics.

We have seen in Eq. (4.3) that the partition coefficient of a solute is related to the ratio of its activity coefficients in the two partitioning phases. The activity coefficients of members of a homologous series are related to their carbon numbers according to the following equation:

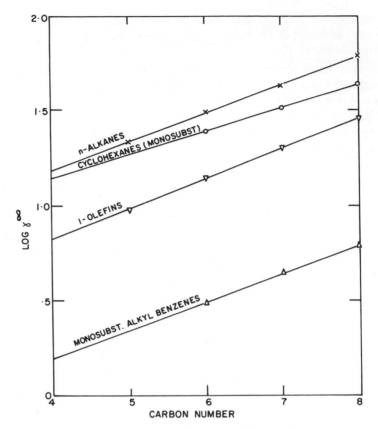

Fig. 4.1. Activity coefficients in acetonitrile by LLC at 25°C.

$$\ln \gamma_i = a + bn_i \tag{4.4}$$

where γ_i is the activity coefficient of the ith member of the series, having n_i carbon atoms, and a and b are constants. Data obtained by Locke[1] from LLC experiments with hydrocarbons in acetonitrile are plotted in Fig. 4.1 and confirm the validity of the equation.

Thus for the stationary and the mobile phases we have, respectively,

$$\ln \gamma_i^{(S)} = a_s + b_s n_i$$
$$\ln \gamma_i^{(M)} = a_M + b_M n_i$$

i.e.,

$$\ln \gamma_i^{(M)} - \ln \gamma_i^{(S)} = \ln K_i = (a_M - a_s) + (b_M - b_s)n_i$$
$$\ln K_i = a' + b'n_i \tag{4.5}$$

We see therefore that the retention of homologues increases exponentially with carbon number; providing b' is large enough, members of any homologous series can be separated from one another by partition chromatography. The requisite magnitude of b', of course, depends on the resolution provided by the chromatographic system; an inefficient column requires a phase pair providing a much larger value of b' than an efficient column.

Most real samples contain compounds of several different types and we can calculate a and b values from Eq. (4.4) for several classes, in both squalane and acetonitrile, from data in Locke's paper[1], as shown in Table 4.1.

Table 4.1

Homologous series	Squalane		Acetonitrile		in LLC	
	a	b	a	b	a'	b'
n-Alkanes	−0.32	0.022	0.57	0.152	0.89	0.130
n-Alk-1-enes	−0.31	0.020	0.17	0.162	0.48	0.142
Alkyl cyclohexanes	−0.33	0.010	0.64	0.124	0.97	0.114
Monosubstituted alkyl benzenes	−0.28	0.021	−0.38	0.145	−0.10	0.124

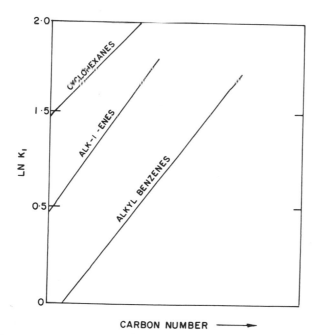

Fig. 4.2. Selectivity of acetonitrile–squalane for hydrocarbon types.

In Fig. 4.2 we have plotted $\ln K_i$ ($= a' + b' n_i$) from these data. From the graph we note that for a given value of $\ln K_i$ (i.e., a given retention volume) the carbon number of the eluting member of the homologous series increases in the order cyclo before n-alkanes before alk-l-enes before alkyl benzenes. Though the lines are not parallel (b' values differ), very roughly the selectivity of the system acetonitrile–squalane relative to cyclohexanes is: n-alkanes retained 0.5 carbon number, alk-l-enes retained 3.5 carbon number, alkyl benzenes retained 8.0 carbon number. In other words, if acetonitrile is used as the mobile phase, n-octyl benzene and nonene-1 will both have been eluted from squalane, at 25 °C, before n-hexane emerges. We would say that acetonitrile–squalane is a very selective separation system for the separation of alkanes and aromatic hydrocarbons.

We have considered the well-documented squalane–acetonitrile system in some detail to get an appreciation of the practical significance of "selectivity." The number of systems for which thermodynamic data are available is very strictly limited and the ability to predict separability or selectivity is normally not a matter of calculation but of empiricism and experience. "Polarity" is a very rough guide to the selectivity which a solvent is likely to show for a series of different compound types and we have seen that selectivity can be deduced from activity coefficient data. There are two other quantitative criteria from which deductions on selectivity may be made, the "solubility parameter" of Hildebrand[2] and the critical solution temperature. These properties, together with activity coefficients and "polarity," are all related thermodynamically and express in their different ways the interaction between different molecular species. Since this book is intended to be a practical guide to liquid chromatography, we will seek merely to show how a knowledge of solubility parameter or critical solution temperatures—for

Table 4.2. Selected Values of Solubility Parameter at 25°C

Perfluoro-n-hexane	5.9
i-Pentane	6.8
n-Pentane	7.1
n-Hexane	7.3
Hexene-1	7.3
Diethyl ether	7.4
n-Hexadecane	8.0
Cyclohexane	8.2
Ethyl chloride	8.3
Carbon tetrachloride	8.6
Ethyl benzene	8.8
Benzene	9.2
Carbon disulfide	10.0
1,2-Dibromoethane	10.2

a From Ref. 9.

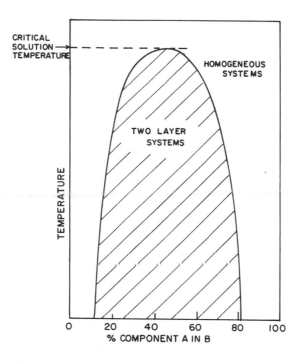

Fig. 4.3. Critical solution temperature of binary system.

both of which there exist reasonably extensive data tabulations([3, 4]) —can be applied to the selection of suitable phase pairs for use in partition chromatography.

The solubility parameter is defined as the square root of "the energy of vaporization per cc" of a pure substance; it can be measured in several ways. The thermodynamic properties of a binary solution can be related to the square of the difference between the solubility parameters of the two components([2]). Some selected values of solubility parameter are given in Table 4.2.

Critical solution temperature T_C is the maximum temperature of a binary liquid system at which two immiscible liquid phases can exist in equilibrium. Figure 4.3 shows the generalized relation between composition of a two-component mixture and temperature; the shaded area represents the range of compositions and temperatures in which two liquid phases are present. Critical solution temperature and solubility parameter are related by the expression

$$2RT_C = f(\delta_A - \delta_B)^2 \qquad (4.6)$$

i.e., two substances having very similar solubility parameters will have a very low T_C and hence will only form two immiscible liquid phases at very low temperatures. A knowledge of solubility parameters is therefore helpful in choosing immiscible phase pairs.

The critical solution temperature gives a guide to selectivity; let us consider values of T_C for systems involving acetonitrile. Some significant values are given in Table 4.3. Looking at the compound group A (alkanes), we can make a very rough extrapolation and conclude that the T_C of squalane (hexamethyltetracosane) is around 170°C, i.e., below 170°C acetonitrile–squalane systems exist as two liquid phases in equilibrium. This deduction is in line with the observation that squalane–acetonitrile is relatively immiscible, a practically usable partition system. Looking at group B, we can make deductions about the selectivity of acetonitrile relative to hydrocarbon types. The alkane and naphthene have very similar T_C values, implying little selectivity between them. The alkene, with a T_C some 40°C lower than the alkane, is significantly more soluble in acetonitrile, while the alkyl benzenes are enormously more soluble. Thus the T_C data indicate a selectivity of acetonitrile increasing in the sequence alkanes $\approx$ naphthenes $<$ alkenes $\ll$ benzenes, which is consistent with the experimental results of Locke which have already been discussed. Thus at this stage we note that T_C data

Table 4.3. **Critical Solution Temperatures of Binary Systems with Acetonitrile**

Second component	T_C. (°C)
A	
n-Pentane	60
n-Octane	91.5
n-Undecane	112.5
2,2,4-Trimethylpentane	81
B	
Methylcyclohexane	78
1-Heptene	38
n-Heptane	84.6
i-Propylbenzene	−78
di-*iso*-Propylbenzene	−15
C	
n-Decanol	22.7
n-Dodecanol	35.2
n-Hexadecanol	58
Methyl myristate	11
Methyl stearate	53.1
Methyl palmitate	31.0
Ethylene glycol	−13.5
Propylene glycol	0
Trimethylene glycol	−6

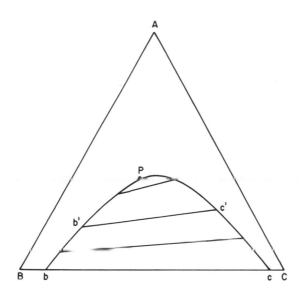

Fig. 4.4. Phase diagram of ternary system with an
immiscible component pair.

can be correlated with experimental results and observations in chromatographic systems. This gives us confidence directionally in predicting the
selectivity of acetonitrile–squalane for other compounds, such as those in
group C of Table 4.3. One would anticipate moderate selectivity of acetonitrile for both alkanols and methyl carboxylates and higher selectivity for
glycols.

The book by Francis[4] contains about 200 pages of tabulated critical
solution temperatures which cover many commonly encountered organic
solvents and, at the practical level, is probably as good a source book for
preliminary screening and prediction of selectivity as exists to date. It also
provides a pointer to the likely degree of miscibility of potential phase pairs
as a function of column temperature.

Discussion of selectivity has so far revolved around two-component
systems, which inevitably, at a fixed temperature, exist as two phases of
definite composition and therefore of invariant selectivity. However, use
of a third component provides a method of great flexibility in generating
immiscible phase pairs of widely differing polarity. Referring to the phase
diagram in Fig. 4.4. components B and C are both completely miscible with
A but only very slightly miscible with one another. A mixture of B and C
will separate into two phases having compositions indicated by b and c. As
the mutually miscible component A is added it distributes between the two
layers and their composition is given by the tieline extremities, such as b′

and c'. The tielines are normally not horizontal, since A will be more soluble in one component than the other. As more A is added, so the compositions of the two phases become more nearly alike and eventually, at P, are identical, i.e., only one liquid phase exists at P (the "plait") point.

In partition chromatography one can profit from the properties of such systems in the following way. B and C might be, for example, water and heptane, almost mutually insoluble and having extremely different polarity characteristics. By addition of an alcohol, such as ethanol, a whole series of two-phase ternary component compositions with graded selectivities can be obtained. In the extreme, just away from the plait point, the two phases will have almost identical composition and therefore almost identical selectivity. In practice to use this system one would have to plot the phase diagram and select the tieline giving a desired selectivity characteristic, and then using material of composition given by one end of the tieline as mobile phase and by the other end for the stationary phase. Adjustment of selectivity could subsequently be obtained *in situ* by addition of one or more of the components to the mobile phase until a new equilibrium was attained. Use of this approach to obtain adjustable selectivity over a wide range has not yet been described in high-performance LC. To an extent it requires a trial-and-error approach in selection of the three components and their proportions since at present it is generally impossible to predict the optimum selectivity required for separation of multicomponent samples. A further disadvantage of this approach lies in its temperature dependence, thus requiring good thermostatting. This, of course, is true of most partition systems and will be discussed next.

4.3.2. Stability of Phase Pairs

The ideal partition system is one in which the stationary liquid phase is in equilibrium with the mobile phase; the amount and distribution of the stationary liquid in the system remain constant. Various effects operate to deny attainment of this ideal, for example, mutual solubilities of the two contacting phases and the effect of changing temperature on solubility.

There are two practical approaches to the attainment of the stable LLC system: equilibration, and the use of chemically bonded phases. By equilibration we mean the presaturation of mobile phase with stationary phase at the column operating temperature. If carried out effectively, the consequence of this is that none of the stationary phase in the column can dissolve in the mobile phase, or, rather, that there is no net transport of stationary phase although exchange of molecules between the phases will, of course, occur. It is a matter of experience that effective pre-equilibration of the phases is less easy to realize than is sometimes imagined. A recommended procedure is as follows: Ensure saturation of the mobile phase by shaking for about 24 hr with excess of the stationary phase at the column temperature. Before using the equilibrated mobile phase it should be allowed to stand for a few hours to allow suspended droplets of the stationary phase to settle out. In addition, in the chromatographic system use a precolumn between the eluent reservoir and the separation column (both pre- and separation column being in the same thermostat) so that any relatively small measure of "under"- or "over"-saturation of the mobile phase can be corrected. The need for this final correction arises because of the likelihood that a small difference in temperature existed between the column and the vessel in which bulk mobile phase had been preequilibrated.

Chemically bonded stationary phases are being used increasingly in chromatography. The idea grew out of the use of silanizing reagents to deactivate solid supports used in gas chromatography by reaction with support surface hydroxylic groups. Abel et al.[5] extended this idea by using highly alkylated silyl chlorides which, after reaction with the surface hydroxyl groups, gave a relatively thick coating of chemically bonded stationary phase. They used this approach to eliminate "bleeding," that is, the loss of stationary phase which occurs in gas chromatography due to its volatility. Stewart and Perry[6] adopted exactly the same approach to prepare stable liquid phases for LLC: they prepared "octadecyl celite," from which organic material was not eluted by any of a range of organic eluents. Subsequently Halasz and Sebestian[7] described similar materials, which are now commercially available under the trade name Durapak and which are sold for use in both gas and liquid partition chromatography: the Zipax materials also include bonded phases of this type (Permaphases)

The structure of bonded phases was proposed by Stewart and Perry[6], following Abel et al.[5], as of the following types:

(I)

$$
\begin{array}{c}
\phantom{nC_{18}H_{37}-Si-O-Si} \diagup\!\!\!O\!\!\!\diagdown \\
\phantom{nC_{18}H_{37}-Si-O-}O\!\!-\!\!Si \diagdown \\
\phantom{nC_{18}H_{37}}\diagup O\!\!\!\diagdown \\
nC_{18}H_{37}\!\!-\!\!Si\!\!-\!\!O\!\!-\!\!Si' \\
\phantom{nC_{18}H_{37}-Si-}\diagdown O \diagdown O \\
\phantom{nC_{18}H_{37}}\diagup O\!\!-\!\!Si \diagup \\
nC_{18}H_{37}\!\!-\!\!Si\!\!\diagdown O\!\!-\!\!Si \diagdown
\end{array}
$$

(II)

At the time of writing three Durapak products are available, in which oxydipropionitrile, Carbowax 400, and n-octane residues are bonded to porous glass. These three materials cover extremes of selectivity and should provide a good basis for useful investigation of liquid partition systems; the major problem which may be anticipated is that the phase-support bond could prove susceptible to hydrolysis, so that use of thoroughly dried eluent may prove to be necessary.

A wide range of Zipax bonded phases is also becoming available and this is clearly an area of rapid commercial exploitation.

4.3.3. Relation of Dynamics to Phase Pair Choice

The properties of stationary and mobile phases which affect chromatographic dynamics and, therefore, performance include mobile phase viscosity and diffusion coefficients in the phases. Of these properties the effect of mobile phase viscosity is the easiest to predict since the values of organic liquid viscosities are readily available and the influence of viscosity on, in particular, analysis time for a given driving pressure is straightforward to determine.

The diffusion coefficients have an influence on column performance, rather than analysis time. High diffusion rates in the stationary phase favor rapid equilibration of solute between the immiscible phases and thus lead to improved resolution, while high diffusion rates in the moving phase have an adverse affect. Unfortunately, data on these properties are limited.

The mobile phase in liquid partition chromatography usually consists of relatively small organic molecules or a mixture of several such materials. This follows from the simple, practical need to use a low-viscosity liquid to minimize the driving pressure needed to force it through a packed bed. Also, for maintenance of degree of separation, time of analysis and eluent viscosity increase together; double the mobile phase viscosity and separation time is doubled. In Table 4.4 viscosities of various organic liquids are recorded, to demonstrate the adverse effect on viscosity of increasing the size and the number of polar groups in organic molecules. The dielectric constants are also tabulated. The dielectric constant of an organic liquid can roughly be equated to the ill-defined property chromatographers call "polarity." As a very crude first approximation, the selectivity of a mobile phase versus an alkane stationary phase increases with increasing polarity. Thus we see from Table 4.4 that as the "selectivity" (dielectric constant) of families of similar organic liquids increases, so, too, does the adverse effect of molecular size and complexity on viscosity. As a general rule Snyder and Saunders[8] imply that it is desirable to use mobile phases with viscosities at working temperature up to about 0.4 cP. Consequently we can, for most practical purposes, limit choice of the mobile phase to a small number of substances. Readily available organic solvents with viscosities up to 0.4 cP at about ambient temperature are listed in Table 4.5, which is extended to also include some other, more viscous, but still useful mobile phases. From the values quoted one can estimate roughly the time or pressure disadvantage their use incurs. The temperature coefficient of viscosity varies quite considerably among the different molecular types; it is particularly large for benzene and reference to the table shows that a 10 °C rise in temperature lowers the viscosity very markedly. Thus it is worth considering the advantages of even modest increases in temperature from the standpoint of faster analyses.

The diffusion coefficients of solutes in liquids are generally about five orders of magnitude smaller than in gases. The significant contribution of plate height made by solute diffusion in the mobile phase in gas chromatography is thus much reduced in liquid chromatography. We discussed in an earlier chapter the magnitude of the contributions to band broadening of the diffusion of solutes in the mobile and stationary phases and will merely recapitulate the conclusions reached (see pp. 22–25). These were that diffusion in the stationary phase was generally much less than in the mobile phase

Table 4.4

Substance	Molecular weight	Viscosity,cP at 20° C	Dielectric constant
Pentane	72	0.23	1.8
Decane	142	0.92	2.0
Methyl acetate	74	0.37	8.5
Butyl acetate	116	0.73	5.01
Butyl stearate	341	~ 8	3.1
Methanol	32	0.60	32.6
Propanol-2	60	2.3	18.3
1,2-Ethane diol	62	19.9	37.7

Table 4.5. Organic Solvents of Suitably Low Viscosity
for Use As Mobile Phase in LLC

Substance	Viscosity, cP at 20°C
n-Pentane	0.23
n-Hexane	0.31
n-Heptane	0.41
Diethyl ether	0.23
i-Propyl ether	0.32
Butanone-2	0.40
Methyl acetate	0.37
Dichloromethane	0.44
i-Propyl chloride	0.33
Acetonitrile	0.37
Carbon disulfide	0.37
Diethylamine	0.38
Benzene	0.65
	0.55[a]
Toluene	0.59
Methanol	0.60
Ethanol	1.20
i-Propanol	2.3
Acetic acid	1.26
Ethyl acetate	0.45
Chloroform	0.57
Nitromethane	0.67
Pyridine	0.94

[a] At 30°C.

(because stationary liquids generally are relatively viscous and of high molecular weight) and its influence could be disregarded in selecting the stationary phase. Choice of the mobile phase merited somewhat more consideration; low plate height contributions are favored by more viscous eluents. This requirement is diametrically opposed to the viscosity–analysis time effect described earlier, but, as the absolute magnitude of the plate height–viscosity effect is small, use of low-viscosity mobile phases is almost always the rule.

In summary, therefore, the most important practical point stemming from consideration of column dynamics is to choose a mobile phase of low viscosity to achieve fast analyses at modest driving pressures.

4.4. SUPPORT MATERIALS

In partition chromatography the stationary liquid phase is supported on a material which is usually of high surface area, small particle size (1–200 μm diameter), and inert.

Substances which meet these requirements have been very extensively studied in connection with gas chromatography and much is known about their structure[9] and the extent to which they may interact with solutes[10] in GC. Diatomaceous earth, kieselguhr, porous glass, porous polymers, adsorbents—such as silica gel and alumina—and cellulose can all be used as supports for the stationary liquid in liquid–liquid partition chromatography. The principal practical difference between supports for liquid and gas chromatography is that for the former much smaller particle sizes are preferred (1–50 μm).

It is worth noting that one advantage LC has over GC is that it enables relatively nonvolatile materials to be separated at ambient temperature. This is a very desirable situation if the sample components are at all labile. There is an incentive therefore to be mindful of possible interaction between supports and the sample: avoidance of pink firebrick (Chromosorb P, Sterchamol, Sil-o-Cel, etc.) materials or silica gel and alumina as supports is recommended in the separation of sensitive material.

New support materials are being developed which are proving particularly valuable. In particular reference must be made to the "controlled-surface-porosity" (CSP) supports available from Du Pont, surface-textured glass beads from Corning Glass Co., and Corasil (Grades I and II) from Waters Associates.

The CSP and Corasil particles consist of solid, spherical siliceous cores with a porous surface of controlled thickness and pore size, as illustrated in Fig. 4.5. The ratio d/t is approximately 30 in CSP supports.

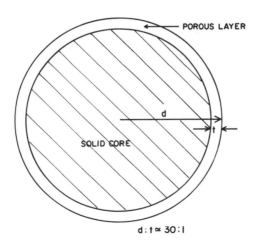

Fig. 4.5. "Porous layer" or "controlled surface porosity" particle.

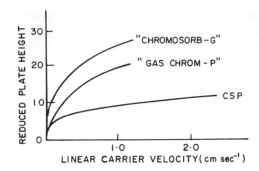

Fig. 4.6. Effect of support on column performance.
Stationary phase, β, β'-oxydipropionitrile.
(Courtesy of *J. Chromatog. Sci.*)

Stationary liquid is retained in the porous coating in exactly the same way as it would be retained in a completely porous particle; however, because of the small depth of stationary liquid retained, important advantages, in terms of column performance, accrue. One of the significant contributory factors to band broadening is the time taken for solutes to diffuse through and out of stationary liquid distributed throughout a porous support. The band broadening becomes more and more serious as the mobile phase velocity increases. One would anticipate that partition columns with CSP supports would give particularly high performance and that their advantages would be especially marked in fast analysis. Data from Kirkland[11] (Fig. 4.6) confirms that the normal deterioration in column efficiency with increasing eluent flow rate is very much reduced using these supports. The ordinate shows the reduced plate height, which is the actual plate height, divided by the average particle diameter. This is a convenient way of reducing plate height data to a common standard. It will be noted that whereas the performance of columns with conventional, completely porous supports (i.e., Chromosorb-*G* and Gas Chrom-*P*) deteriorates sharply as the carrier (mobile phase) velocity increases, the deterioration with CSP supports is strikingly reduced.

4.5. FURTHER PRACTICAL CONSIDERATIONS

4.5.1. Preparation of Stationary Phases

Until comparatively recently stationary phases for liquid partition chromatography have consisted of liquids dispersed over the surface of a porous particle. The methods of preparing the material in a form suitable for packing into columns are straightforward and identical with those familiar in gas chromatography. The stationary liquid (a known quantity) is dissolved in a relatively volatile solvent; the solution is added to the appropriate amount of the porous support, which has been "moistened" with the same solvent. The solvent is allowed to evaporate, the slurry of support, solvent, and stationary liquid being agitated during the evaporation to facilitate uniform dispersion of the liquid over the support. Last traces of the solvent are removed by an appropriate combination of heat and/or vacuum treatment; the finished material should be completely free-flowing. The above recipe is by no means uniquely suitable; it is offered as a guide to those with no previous experience of stationary phase preparation.

The chemically bonded stationary phases are much more difficult to prepare. Typically the surface hydroxyl groups of a porous support or adsorbent are reacted with substituted chlorosilanes, as described in Ref.[11] However, it is strongly recommended that commercial bonded phases should be purchased; despite the high cost of these products, their convenience, and the time and, therefore, cost of "doing-it-yourself" rule out the latter approach except perhaps in academic circles.

4.5.2. Liquid Phase "Loading" and Sample Capacity

By "loading" is meant the amount of stationary liquid phase deposited on unit quantity of support. Supports like the various diatomaceous earths will take up as much as one-half their own weight of liquid without becoming "tacky" and the usual range of composition of stationary phases of the conventional, nonbonded type is 1–40 parts of stationary liquid to 100 parts of support. Chemically bonded materials may have up to about 25% of organic groupings attached to the support, but where solid-cored particles are used this proportion drops to a very much lower level, around 0.5–1%. This level is also that used with nonbonded liquids on these same solid-cored supports.

The amount of liquid in the stationary phase controls the maximum amount of a sample which can effectively be separated on that phase. Crudely, the greater the amount of liquid, the larger is the sample that can be handled. In high-performance column partition chromatography, sample sizes are of the same order as in gas chromatography; for heavily loaded columns, over 20% of stationary liquid, samples of up to 10 mg are usable, whereas for the lightly coated materials such as the porous layer beads (CSP) it is good practice to use quantities only in the range up to 1 mg. Asymmetric peaks, decreased resolving power, and less reproducible retention volumes are the practical disadvantages which arise from using overlarge samples.

REFERENCES

1. Locke, D. C., *J. Chromatog.* **35,** 24 (1968).
2. Hildebrand, J. H., and Scott, R. L., *Regular Solutions,* Prentice-Hall, Englewood Cliffs, New Jersey (1962).
3. *Solubility of Non-Electrolytes,* Dover, New York p. 435 *et seq.,* (1964).
4. Francis, A. W., *Critical Solution Temperatures,* American Chemical Society, Washington (1961).
5. Abel, F. W., Pollard, F. H., Uden, P. C., and Nickless, G., *J. Chromatog.* **22,** 23 (1966).
6. Stewart, H. N. M., and Perry, S. G., *J. Chromatog.* **37,** 97 (1968).
7. Halasz, I., and Sebestian, I., *Angew Chem., Intern. Ed. (English)* **8,** 453 (1968).
8. Snyder, L. R., and Saunders, D. L., *J. Chromatog. Sci.* **7,** 195 (1969).
9. Ottenstein, D., *Advan. Chromatog.* **3,** 137 (1966).
10. Kusy, V., *Anal. Chem.* **37,** 1748 (1965).
11. Kirkland, J. J., *J. Chromatog. Sci.* **7,** 2, 361 (1969).

FURTHER READING

Williamson, A. G., *An Introduction to Non-Electrolyte Solutions,* Oliver and Boyd, Edinburgh, (1967). (Presents a good general and not too detailed account of the physical chemistry of nonelectrolyte solutions.)
Keller, R. A., and Giddings, J. A., Theoretical basis of partition chromatography, *Chromatography,* Heftmann, E., ed., Reinhold, New York, 2nd ed. (1967). (Gives a general introduction to the topic, with many references.)

Chapter 5

Exclusion Chromatography—Mechanism and Materials

5.1. INTRODUCTION

Exclusion chromatography is a branch of liquid chromatography in which a solute is distributed between the free solvent and the solvent contained in the interstices of porous particles. The free solvent is the mobile phase and the porous particles containing the solvent constitute the stationary phase (the porous particles themselves are often loosely referred to as the stationary phase). The separation depends upon the fact that the interstices are of such a size that the degree of penetration of the solute molecules depends upon the size of these molecules. Very large molecules are unable to enter the stationary phase and are eluted in a volume of solvent equal to the void volume of the column, i.e., the volume of the free solvent in the column. If the solute molecules are sufficiently small so that they can penetrate into the whole of the solvent contained in the stationary phase, they are eluted by a volume of solvent equal to the void volume of the column plus the volume of solvent contained in the stationary phase, i.e., equal to the total volume of solvent in the column. Thus, providing adsorption effects are absent, exclusion chromatography separates according to size, large molecules being eluted before small molecules. All solutes are eluted in the comparatively narrow range of eluent volumes between the void volume of the column and the total volume of the solvent in the column.

Between the two extremes the elution volume is governed by the same relation as that found for partition chromatography, i.e.,

$$V_r = V_o + KV_s \qquad (5.1)$$

where V_r is the elution volume of the solute, V_o is the void volume of the

Table 5.1. Maximum Number of Peaks in Exclusion, Gas, and Liquid
Chromatography

Theoretical plates	Peak capacity		
	Exclusion	Gas	Liquid
100	3	11	7
400	5	21	13
1,000	7	33	20
2,500	11	51	31
10,000	21	101	61

column, V_s is the volume of the pores in the stationary phase, and K is the distribution coefficient of the solute between the free solvent and the solvent in the stationary phase. Thus KV_s can be regarded as the volume of solvent in the stationary phase freely available to the solute. If the solvent in the stationary phase is completely permeable to the solute, $K = 1$, and if completely impermeable, $K = 0$.

This limited range of distribution coefficients is unique to exclusion chromatography. It means that all components are eluted between volumes V_o and $V_o + V_s$ of the solvent. Obviously the number of components that can be separated will be related to the efficiency of the columns. In other forms of chromatography K can exceed 1 and this restriction does not apply. Giddings[1] has calculated the peak capacity of exclusion, gas, and liquid chromatography columns with a given number of theoretical plates; his results are shown in Table 5.1.

Thus in exclusion chromatography long efficient columns are necessary for good separations. The fact that K cannot exceed 1 has the advantage in that when an elution volume of $V_o + V_s$ is reached, no material remains on the column and it is possible to separate more than one sample at a time on the same column providing the sample injection times are arranged to avoid overlap.

Much of the early work in exclusion chromatography was carried out in aqueous solutions using cross-linked dextrans as the stationary phase. When these gels are swollen they are relatively soft and most of the work was carried out in glass columns with low solvent flow rates. The main applications were the separation of substances of biochemical interest and also in the determination of molecular weights. The latter application was particularly successful in the case of globular proteins. In this chapter we shall be concerned with nonaqueous usage.

The stationary phases used with aqueous solutions could not be used with organic solvents since they deswelled and became impermeable. It was found by Moore[2] that porous polystyrene–divinyl benzene gels could be used with a wide variety of organic solvents. Moore developed an apparatus using a refractive index detector which was well suited to the fractionation of commercial polymers and it became widely used in industry.

5.2. PEAK BROADENING IN EXCLUSION CHROMATOGRAPHY

Providing the solutions injected onto the columns are sufficiently dilute, the peaks obtained are approximately Gaussian and the random walk theory described in Chapter 2 is applicable with some change in emphasis. As with LLC, longitudinal diffusion is negligible. As the molecular weight of the solute increases, the resistance to mass transfer increases; at a certain molecular weight it reaches a maximum and begins to decrease, finally becoming zero when the solute molecules are too large to penetrate the stationary phase. From this point, peak broadening is due solely to eddy diffusion. This is seen in Fig. 5.1; the polystyrene with a molecular weight of 860,000 is totally excluded from the Styragel packing and is eluted at the void volume of the column[3]. At low mobile phase velocities the mass transfer term for large molecules makes a greater contribution to plate height than small molecules, but in the former case the effect of increasing the mobile phase velocity is less pronounced. As the random walk theory predicts, peak broadening due to eddy diffusion decreases as the particle size of the stationary phase decreases; increasing the size range has little effect. Conditions for minimizing peak broadening are the same as for other forms of chromatography: narrow columns, small stationary phase particles, and low-viscosity

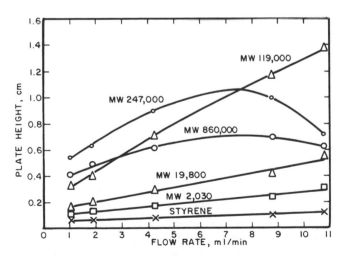

Fig. 5.1. Variation of plate height with molecular weight and flow rate for styrene and polystyrenes[3]. Stationary phase, 10^4 Å Styragel; mobile phase, chloroform. (Reproduced by courtesy of the American Chemical Society.)

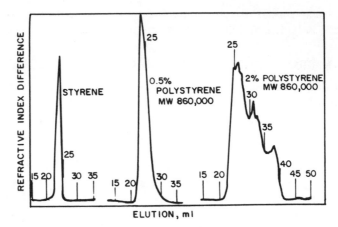

Fig. 5.2. Variation of dispersion with molecular weight and concentration[3]. Stationary phase, smooth glass beads; mobile phase, toluene. (Reproduced by courtesy of American Chemical Society.)

solvents. The plate height increases with solvent velocity, but at least for moderate increases in velocity, the effect is not large.

The concentration of the solution injected onto the column is more critical if the solute molecular weight is high. It would not be unreasonable to inject a 5% solution of a solute with a molecular weight of 500, but if the same concentration were used for a polymer with a high molecular weight, the elution peak would be grossly distorted as shown in Fig. 5.2[3]. This is due to the high viscosity of the polymer solution, which increases the resistance to mass transfer and also the broadening due to eddy diffusion. A concentration of 0.1–0.5 wt% would be more appropriate in the case of the high-molecular-weight polymer. The injection time is less critical than the concentration, but it should be kept as small as possible.

5.3. THEORIES OF EXCLUSION CHROMATOGRAPHY

The elution volume of a molecule depends on the fraction of the time it spends in the stationary phase. This will be governed by the probability of the molecule diffusing into a pore, which will depend mainly on the size of the pore and the size of the molecule. It can be shown that, providing the diffusion coefficient of the molecule in the stationary phase is not greatly different from that in the mobile phase, the time taken for the molecule to diffuse into and out of a particle of the stationary phase is small compared with the time it takes for the solute band to pass the particle. The separation can therefore be treated as an equilibrium process.

Some of the first stationary phases to be used swell appreciably when immersed in the chromatographic solvent, e.g., cross-linked dextrans in water and rubbers in various organic solvents. This is because the osmotic pressure generated by the polymer molecules is balanced by the elastic retraction of the network. High pressures can be generated in the gels and it can be shown that this pressure will decrease the solubility of solute molecules and also that, as the size of the solute molecules increase, so their solubility decreases. According to this theory the solute is partitioned between the mobile phase and the mobile phase under pressure. Ginsberg and Cohen[4] suggested that osmotic pressure was responsible for the exclusion of nonelectrolytes from cross-linked dextrans. It would seem, however, that osmotic pressure can only account for a small part of the exclusion effect, although it is certainly operative in swollen gels.

The fact that the volume of the solvent in the stationary phase available to the solute decreases as the molecular weight increases led early workers to propose simple geometric theories of exclusion. One of the first models proposed was that due to Porath[5]. He assumed that the pores in the stationary phase were conical in shape so that the depth to which a solute molecule could penetrate was smaller, the greater its radius. A more realistic model was proposed by Laurent and Killander[6], who assumed that the polymer chains constituting the matrix of the stationary phase were rigid rods distributed at random. They applied a formula developed by Ogston[7], to show how the distribution coefficient of a spherical molecule varied with its radius and also with the concentration of the rods and their radius. Both the Porath and Laurent–Killander models were supported by experimental evidence, but the tests applied were probably not very sensitive. From these theories it would be expected that if the molecules were smaller than the pores, no separation would be obtained, but Moore and Arrington[8] and de Vries et al.[9] have found that this is not the case for flexible molecules.

The most satisfactory theory propounded so far has been one based on thermodynamic arguments by Casassa[10] for flexible chain molecules and Giddings et al.[11] for globular molecules. The configurations of flexible chain molecules are continually changing but they have a certain statistical average size. Suppose such a molecule enters a pore; if the pore is not too large, the molecule must suffer some loss of configurational entropy. It can be shown that the accompanying enthalpy change is negligible, so that the change in free energy, and hence the distribution coefficient, is determined by the entropy change. Casassa showed that the distribution coefficient is determined only by the dimensions of the molecule and the size of the pore. He was able to calculate the distribution coefficient of polystyrenes in toluene-porous glass and obtained good agreement with chromatographic data. Figure 5.3, due to Yau et al.[3], is of interest in this connection. The abscissa is the distribution coefficient K of a solute calculated by Casassa's theory as a function of the radius of gyration R and the size of the pore a for pores

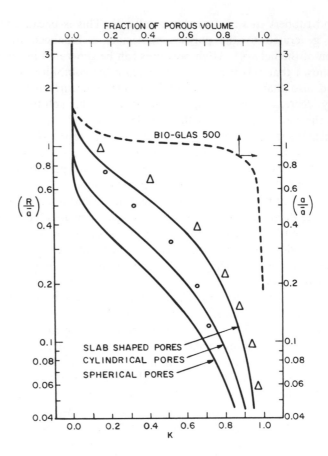

Fig. 5.3. Comparison of column data with pore size distribution and with Casassa's theoretical curves. ($\triangle$) elution points for Bio-Glas 500; ($\bigcirc$) elution points for Bio-Glas 200[3]. (Reproduced by courtesy of the American Chemical Society.)

of various shapes. The solid lines show the theoretical curves predicted by the theory. The chromatographic results for columns packed with Bio-Glas 500 and Bio-Glas 200 are also plotted. The dashed curve is the pore size distribution of Bio-Glas 500.

Casassa and Tagami[12] have shown that the equilibrium theory is applicable even at high flow rates providing that a given molecule undergoes a large number of transfers between the two phases during its passage down the column. This explains the fact that variation in flow rates have little effect on peak elution volumes.

5.4. STATIONARY PHASES FOR EXCLUSION CHROMATOGRAPHY

The remaining part of this chapter is devoted to describing the properties of stationary phases for exclusion chromatography in nonaqueous solutions; its purpose is to give the reader some idea of the type of materials commercially available. These stationary phases can be divided into two groups depending on their composition. Porous silica and glass are inorganic polymers, whereas alkylated cross-linked dextrans, polyvinyl acetates, and polystyrenes have an organic matrix. These organic polymers are cross-linked and therefore insoluble in all solvents provided they are not degraded. However, they swell in certain solvents; the degree of swelling depends on the solvent, the degree of cross-linking, and the method of preparation.

An important property of a stationary phase is that it should give an adequate separation over the molecular weight range of interest. Often the only information available is the exclusion limit; in this case it can be assumed that the maximum separating power will occur over 1-1-1/2 decades of molecular weight below this figure; even the most porous materials give some separation of low-molecular-weight compounds. All molecules with weights above the exclusion limit are eluted together without separation. For small molecules satisfactory separations can be obtained by the use of a single stationary phase with a low porosity, but for high polymers, which may contain molecules with molecular weights from 10^3 to 10^6, several columns connected in series, each packed with gels of varying porosity, are necessary to give an approximately linear log (molecular weight)–elution volume curve over this range of molecular weights. It should be noted that the total number of theoretical plates in such a combination is not necessarily the sum of those of the individual columns[13]. If the columns are approximately equivalent, i.e., the column lengths and elution volumes are nearly identical, the total number of theoretical plates N of n columns joined in series is given by

$$N = n^2 / \Sigma(1/N_i) \tag{5.2}$$

where N_i is the number of theoretical plates of the ith column.

The stationary phase must also be obtainable in the optimum size range: the narrower this range for a given mean diameter, the smaller the pressure drop in the column. It must also be stable so that frequent calibration is unnecessary. Mechanical strength and thermal stability are important. The degree of swelling should also be small since soft materials may bed down during use with alteration of the flow characteristics of the column. In addition, packing columns is simpler with nonswelling materials and the solvent can be changed without greatly affecting column efficiencies. When separations based solely on size are desired, adsorption should be absent; although this effect can often be minimized by modifying the surface or using a polar solvent, it is obviously desirable to choose a material in which these effects are small. The role of the solvent is less important than in other forms of chromatography. That chosen will depend mainly on the solubility characteristics of the sample and the type of detector used. The solvent temperature will be mainly governed by the solubility of the sample; some polymers are only soluble at elevated temperatures, e.g., polyethylene and polypropylene are normally run at 130°C.

Table 5.2. Commercially Available Stationary Phases for Exclusion Chromatography

Stationary phase	Pore diameter, Å	Mol. wt. exclusion limit	Mol. wt. operating range	Supplier[a]
Porous glass				
Bio-Glas 200	200	30,000[b]	3,000–30,000[b]	1
Bio-Glas 500	500	100,000[b]	10,000–100,000[b]	1
Bio-Glas 1000	1000	500,000[b]	50,000–500,000[b]	1
Bio-Glas 1500	1500	2,000,000[b]	400,000–2,000,000[b]	1
Bio-Glas 2500	2500	9,000,000[b]	800,000–9,000,000[b]	
CPG-10-75	75	28,000[c]	—	2,3
CPG–10–125	125	48,000[c]	—	2,3
CPG–10–175	175	68,000[c]	—	2,3
CPG–10–240	240	120,000[c]	—	2,3
CPG–10–370	370	400,000[c]	—	2,3
CPG–10–700	700	1,200,000[c]	—	2,3
CPG–10–1250	1250	4,000,000[c]	—	2,3
CPG–10–2000	2000	12,000,000[c]	—	2,3
Porous silica				
Porasil 60	100	60.000	500–60,000	3
Porasil 250	100–200	250,000	5,000–200,000	3
Porasil 400	200–400	400,000	5,000–400,000	3
Porasil 1000	400–800	1,000,000	10,000–1,000,000	3
Porasil 1500	800–1500	1,500,000	100,000–1,500,000	3
Porasil 2000	1500	2,000,000	500,000–	3
Merck–O–gel Si 150	120–220	50,000[d]	—	4
Merck–O–gel Si 500	300–700	400,000[d]	—	4
Merck–O–gel Si 1000	700–1300	1,000,000[d]	—	4
Alkylated cross-linked dextrans				
Sephadex LH-20	—	—	—	5
Polyvinyl acetate gels				
Merck–O–gel OR 750		750[e]	—	4
Merck–O–gel OR 1500		1,500[e]	—	4
Merck–O–gel OR 5000		5,000[e]	—	4
Merck–O–gel OR 20,000		20,000[e]	—	4

Table 5.2 *(Continued)*

Stationary phase	Pore diameter, Å	Mol. wt. exclusion limit	Mol. wt. operating range	Supplier[a]
Polystyrene divinylbenzene gels				
Styragel 39720	60[b]	—		3
Styragel 39721	100[b]	—		3
Styragel 39722	250[b]	—		3
Styragel 39723	500[b]	—		3
Styragel 39724	1,000[b]	—		3
Styragel 39725	3,000[b]	—		3
Styragel 39726	10,000[b]	—		3
Styragel 39727	30,000[b]	—		3
Styragel 39728	100,000[b]	—		3
Styragel 39729	300,000[b]	—		3
Styragel 39730	1,000,000[b]	—		3
Styragel 39731	10,000,000[b]	—		3
Bio Beads S–X1	14,000[b]	600–14,000		1
Bio Beads S–X2	2,700[b]	100–2,700		1
Bio Beads S–X3	2,000[b]	Up to 2,000		1
Bio Beads S–X4	1,400[b]	Up to 1,400		1
Bio Beads S–X8	1,000[b]	Up to 1,000		1

[a] Suppliers: 1. Bio-Rad Laboratories, 32nd & Griffin Avenue, Richmond, Calif. 94804. 2. Chromatography Products, Corning Glass Works, Corning, N. Y. 14830. 3. Waters Associates Inc., 61 Fountain Street, Framingham, Mass. 01701. 4. E. Merck and A.G. Darmstadt, W. Germany. 5. Pharmacia Fine Chemicals, Uppsala, Sweden.
[b] For polystyrenes in toluene.
[c] For aqueous solution of dextrans.
[d] For polystyrenes in chloroform.
[e] For polystyrenes in tetrahydrofuran.

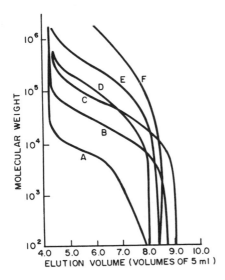

Fig. 5.4. Elution volume–molecular weight relationship for various grades of Porasil; solutes, polystyrenes; mobile phase, toluene[17]. A, Porasil 100 Å pore diameter; B, Porasil 100–200 Å pore diameter; C, Porasil 200–400 Å pore diameter; D, Porasil 400–800 Å pore diameter; E, Porasil 800–1500 Å pore diameter; F, Porasil 1500 Å pore diameter. (Reproduced by courtesy of the American Chemical Society.)

Table 5.3. Polymer Standards for Exclusion Chromatography

Polystyrene			
Nominal molecular weight		M_w/M_n	Supplier[a]
600		1.10	1,4
2,100		1.10	1,4
4,000		1.10	1
10,000		1.06	1,4
20,000		1.06	1,4
51,000		1.06	1
97,200		1.06	1,4
200,000		1.06	1,4
498,000		1.20	1,4
670,000		1.15	1,4
1,800,000		1.20	1,4
179,000		1.05	3

Polyethylene			
Peak molecular weight		M_w/M_n	Supplier
7,100		1.7	1
17,900		1.95	1
27,700		2.0	1
32,000		6.0	1
35,700		4.2	1
41,700		3.7	1

Polyvinyl Chloride			
M_w	M_n	M_w/M_n	Supplier
68,000	25,500	2.61	1,4
118,200	41,000	2.88	1,4
132,000	54,000	2.44	1,4

Linear Polybutadiene			
M_w	M_n	M_w/M_n	Supplier
17,000	16,100	1.56	2
170,000	135,000	1.26	2
272,000	206,000	1.32	2
332,000	226,000	1.47	2
432,000	286,000	1.51	2

Linear Hydrogenated Polybutadiene			
M_w	M_n	M_w/M_n	Supplier
108,000	82,000	1.32	2
164,000	141,000	1.16	2
194,000	126,000	1.54	2
420,000	158,000	2.66	2

[a] Suppliers: 1. Pressure Chemical Co., 3419 Smallman St., Pittsburgh, Pa. 15201. 2. Phillips Petroleum Co., P.O. Box 968, Phillips, Tex. 79071. 3. National Bureau of Standards, Washington, D.C. 20234. 4. Arro Laboratories Inc., P.O. Box 686, Joliet, Ill. 60434.

The properties of some commercially available stationary phases are shown in Table 5.2. Some commercially available polymer standards are shown in Table 5.3.

5.4.1. Inorganic Stationary Phases

When porous glass and silica stationary phases were first introduced it was thought that they would replace organic gels for the fractionation of high-molecular-weight polymers. They were available in a range of closely controlled pore sizes covering the high-molecular-weight ranges. The fact that they did not swell meant that packing columns was simple and that solvents could be changed without greatly affecting column efficiencies; they were also stable at high temperatures. Unfortunately, the HETP of columns packed with these materials is usually higher than those packed with polystyrene–divinyl benzene gels. In addition they often show undesirable adsorption effects and have to be deactivated before use. Porous glass packing was introduced by Haller[14] and its use evaluated by him in a subsequent paper[15] and also by Moore and Arrington[8]. An account of the properties of porous silica will be found in an article by de Vries et al.[16]. Figure 5.4

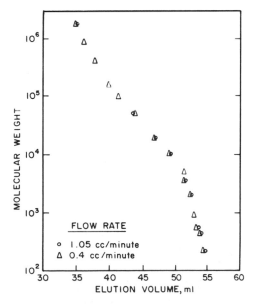

Fig. 5.5. Elution volume–molecular weight relationship for broad pore-size-distribution porous glass; solutes, polystyrenes; mobile phase, toluene[18]. (Reproduced by courtesy of the American Chemical Society.)

shows calibration curves for 4-ft columns of various grades of 105–125-μm Porasil[17]. A similar curve for a 64.5-in. column packed with 125–150-μm porous glass with a wide-pore size distribution (Bio-Glas BRX 85001) and treated with hexamethyldisilazane is shown in Fig. 5.5. A very similar curve was obtained by a combination of columns containing the narrower pore-sized-distribution materials Bio-Glas 200, 500, 1000, and 1500[18].

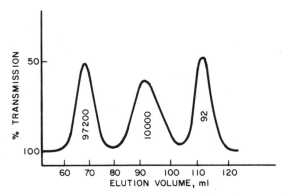

Fig. 5.6. Separation of toluene and polystyrenes on Merckogel Si 500; mobile phase, chloroform[19]. [Reproduced by courtesy of Waters Associates (Instruments) Ltd.]

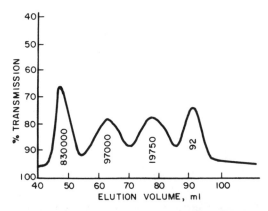

Fig. 5.7. Separation of toluene and polystyrenes on Merckogel Si 500; mobile phase, chloroform[19]. [Reproduced by courtesy of Waters Associates (Instruments) Ltd.]

Separations of polystyrene obtained with various Merck large-pore-size silica gels are shown in Figs. 5.6–5.9. The 1m columns were packed with 50–200-μm particles and had an efficiency of about 1000 theoretical plates per meter: chloroform was used as the mobile phase with a UV detector. A good separation was also obtained with benzene, but tailing occurred with tetrahydrofuran. Some polymers were irreversibly adsorbed on the silica gels.

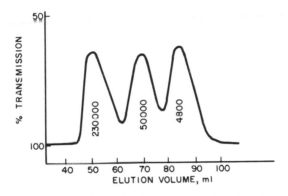

Fig. 5.8. Separation of polystyrenes on Merckogel Si 1000. Mobile phase, chloroform[19]. [Reproduced by courtesy of Waters Associates (Instruments) Ltd.]

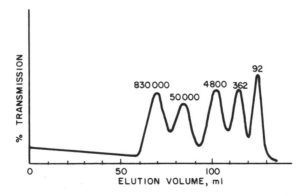

Fig. 5.9. Separation of toluene, polystyrenes, and oligo-phenylenes on Merckogel 1,000,000; mobile phase, tetrahydrofuran[19]. [Reproduced by courtesy of Waters Associates (Instruments) Ltd.]

5.4.2. Alkylated Cross-Linked Dextrans

In most of the work published dealing with separations in aqueous solutions cross-linked dextrans commercially available as Sephadex have been used as the stationary phase. If most of the hydroxyl groups are alkylated, Sephadex becomes more lipophilic and will swell in some polar solvents. The product obtained by alkylating Sephadex G25 is known as Sephadex LH20. The exclusion limit is less than 5000, so that it is only suitable for fractionating low-molecular-weight material. Sephadex LH20 exhibits aromatic adsorption properties, and substances containing hydroxyl and carboxyl groups are adsorbed when chloroform is used as the mobile phase. The adsorption properties have been used in some practical applications, e.g., separation of polynuclear aromatics and of cycloparaffins from paraffins[20]. For the separation of small molecules based on size, the adsorptive properties of this material can be a nuisance and the use of polystyrene-divinyl benzene gels with a low exclusion limit would be preferable.

5.4.3. Polyvinyl Acetate Gels

These lipophilic gels, sold under the name Merckogel OR, are produced by the copolymerization of vinyl acetate and butanediol-1,4 divinyl ether or divinyl esters of dicarboxylic acids[19]. Up to exclusion limits of 5000 they are homogeneously cross-linked; the more porous gels are prepared by polymerization in the presence of an inert diluent. Even the gels with high exclusion limits swell quite considerably in organic solvents and this may be responsible for the linear log(molecular weight) versus elution volume curve (Fig. 5.10). When swollen the gels are rigid and can be used in columns with or without the application of pressure.

5.4.4. Polystyrene–Divinyl Benzene Gels

Gels made by cross-linking polystyrene with divinyl benzene have been investigated by Green and Vaughan[21], who showed that some fractiona-

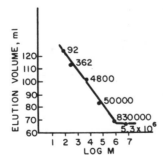

Fig. 5.10. Elution volume versus log (molecular weight) for Merckogel OR 1,000,000; solutes, toluene, polystyrenes, and oligophenylenes[19]. [Reproduced by courtesy of Waters Associates (Instruments) Ltd]

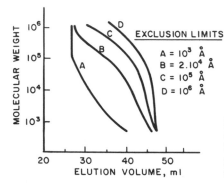

Fig. 5.11. Relation between molecular weight and elution volume for styrene–divinyl benzene gels; solutes, polystyrenes; mobile phase, tetrahdrofuran[22]. (Reproduced courtesy of John Wiley and Sons, Inc.)

tion of polystyrene could be obtained with this material. If the porosity of the gels was increased by decreasing the divinyl benzene content, they swelled considerably and became too soft for use in columns. It was found by workers in the ion-exchange field that it was possible to prepare rigid, but porous polystyrene –divinyl benzene gels if the polymerization was carried out in the presence of certain diluents. If the diluent was a nonsolvent for the polymer, a very porous gel resulted, whereas a good solvent such as toluene produced a much less porous material. By varying the proportion of solvent to nonsolvent it was possible to prepare a whole range of gels with graduations in porosity. As a high proportion of divinyl benzene was used, the gels were rigid and swelled little in organic solvents. The gels are prepared by suspension polymerization, giving approximately spherical particles and are available in various size ranges. They can be used at quite high pressures without compacting and are stable up to about 150 °C in the absence of air. They exhibit practically no adsorptive properties. To obtain high column efficiencies, columns must be packed under high pressures; both the calibration curve and the column efficiency are likely to change if the mobile phase is changed.

A wide variety of organic solvents can be used; the one selected is usually dictated by the solubility of the sample and the type of detector employed. With a refractive index detector the difference in refractive index between the sample and the solvent should be as large as large as possible. For operations at room temperature tetrahydrofuran is often used as it is an excellent solvent and has a fairly low refractive index. O-Dichlorobenzene or 1,2,4-trichlorobenzene are often used as solvents at high temperatures; they have a higher refractive index than polyethylene or polypropylene.

The commercially available gels are usually classified in terms of their upper permeability limits in angstrom units; this can be translated into ordinary molecular weight units for polystyrene by multiplying by 40. The least porous gel has a permeability limit of 45 Å, and is suitable for sepa-

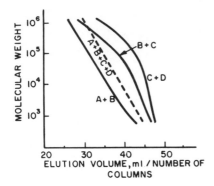

Fig. 5.12. Relation between molecular weight and elution volume for combinations of columns shown in Fig. 5.11[22]. (Reproduced by courtesy of John Wiley and Sons, Inc.)

rating substances with molecular weights of less than about 1000. Figures 5.11 and 5.12, due to Meyerhoff[22], show the relation between elution volume and molecular weight for columns packed with polystyrene–divinyl benzene gels of various porosities both singly and in combination. It will be seen that the four column combination gives an approximately linear log (molecular weight) versus elution volume plot over a wide molecular weight range.

APPENDIX: APPLICATIONS OF EXCLUSION CHROMATOGRAPHY

(1) High-Molecular-Weight Compounds

It is convenient to divide the applications of exclusion chromatography into two parts, depending on whether the solute is a low- or high-molecular-weight substance. Before discussing high-molecular-weight compounds the relation between viscosity and molecular weight will be considered.

If η is the viscosity of a dilute polymer solution and η_0 is the viscosity of the solvent, the specific viscosity η_{sp} is $(\eta - \eta_0)/\eta_0$. If c is the concentration of the polymer solution, the intrinsic viscosity $[\eta]$ is defined as the limit of η_{ps}/c as c tends to zero.

The relation between viscosity and molecular weight is given by the empirical Mark–Houwink equation

$$[\eta] = kM^a \tag{5.3}$$

where k and a are constants for a given solvent–polymer system at a given temperature and M is the viscosity average molecular weight. Values of k and a must be determined experimentally; a compilation will be found in Ref. 23.

Einstein derived the following equation for the viscosity of a dilute suspension of rigid spheres:

$$\mu_{sp} / \mu_0 = 1 + 2.5\phi \qquad (5.4)$$

where ϕ is the fraction of the total volume occupied by the spheres. From this equation it can be shown that for polymers;

$$[\eta] \propto v/M \qquad (5.5)$$

where M is the molecular weight and v is the volume of the equivalent sphere, i.e., a sphere whose volume would enhance the solvent viscosity in a similar manner to the actual polymer molecule.

The number ,weight, and viscosity average molecular weights are defined as follows:

The number average $\bar{M}_n = (\Sigma N_i M_i) / \Sigma N_i$.
The weight average $\bar{M}_w = (\Sigma N_i M_i^2) / \Sigma N_i M_i$.
The viscosity average $\bar{M}_v = [(\Sigma N_i M_i^{1+a}) / \Sigma N_i M_i]^{1/a}$

Here N_i is the number of molecules of molecular weight M_i and a is the exponent in the Mark–Houwink equation. The value of $\bar{M}_v$ is nearer to $\bar{M}_w$ than $\bar{M}_n$ and is identical to $\bar{M}_w$ if $a = 1$.

Prior to the introduction of exclusion chromatography the determination of the molecular weight distribution of polymers was a lengthy process usually carried out by fractional solution or precipitation. Suitable experimental conditions had to be found for each type of polymer and the molecular weights of the fractions also had to be determined. Exclusion chromatography enables this information to be obtained much more rapidly and is applicable to any polymer provided it is soluble. It should be noted that exclusion chromatography is a method of separation and by itself gives no information about molecular weights; it must ultimately be standardized against samples whose molecular weights have been determined by absolute methods. In many cases sufficient information can be obtained by comparing the chromatograms of samples, but if the molecular weight distribution is required, the instrument must be calibrated, i.e., it is necessary to relate the response of the detector to concentration and the elution volume to molecular weight. Differential refractometers have been almost invariably used as detectors since above a molecular weight of about 10,000 the refractive index does not change with moecular weight and output is directly proportional to concentration. For samples containing substantial amounts of material with a molecular weight of below 10,000 the differential refractometer must be calibrated. The elution volume can be related to molecular weight in several ways:

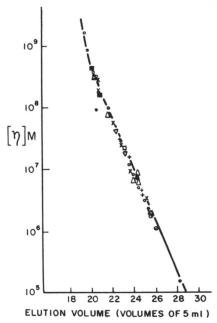

Fig. 5.13. Elution volume versus log $(M[\eta])$ for various polymers. Columns: 10^6, 10^5, 10^4, and 2×10^2 styrene–divinyl benzene gels; mobile phase: tetrahydrofuran([24]). ●, polystyrene; ○, polystyrene "comb"; +, polystyrene "star"; △, hetero graft copol; ×, polymethylmethacrylate; ⊙, polyvinylchloride; ▽, graft copol PS/PMMA; ■, polyphenylsiloxane; □, polybutadiene. (Reproduced by courtesy of John Wiley and Sons, Inc.)

(a) The polymer can be separated into fractions with narrow molecular weight distributions and determination of the molecular weights made by conventional methods. The elution volumes of the fractions are then determined. As a separate calibration curve must be established for each different type of polymer, it is only worthwhile using this method when many samples of a few polymers have to be examined.

(b) The universal calibration method can be used. Benoit et al.([24]), suggested that the important parameter in determining the elution volume of a molecule is its hydrodynamic volume. By Eq. (5.5) this is seen to be proportional to $[\eta]M$. Benoit et al. and subsequent workers have shown that many different types of polymers fall on the same plot of $\log(M[\eta])$ versus elution volume, as shown in Fig. 5.13. In using this method of calibration the coefficients in the Mark–Houwink equation must be known. A review of this method will be found in the paper by Coll and Gilding([25]).

(c) It was originally thought that the elution volume of a polymer depended on the extended chain length of the polymer molecule, i.e., the length of the polymer molecule when fully extended as determined by the valence angles and bond lengths. Once the relation between the logarithm of the extended chain length and elution volume had been obtained for one polymer it could be applied to other polymers by multiplying the extended chain length by the theoretical value of molecular weight per unit chain length, usually known as the Q factor. This method seemed to be particularly attractive since the calibration curve could be obtained from well-characterized samples of polystyrene with narrow molecular weight distributions, which were available. It is now known that the calculated Q values can be in considerable error, but the method can still be applied if the Q value is determined experimentally by comparing either the number, weight, or viscosity average molecular weight calculated in terms of extended chain length from the distribution with that obtained experimentally by osmometry, light scattering, or viscometry. Factors for calculating the extended chain length are given in a paper by Hendrickson[26]. This method is convenient when a large number of different types of samples have to be dealt with; details are given in Ref. 27.

The elution curves obtained in exclusion chromatography are distorted by the peak broadening. This distortion decreases as column efficiency increases and is less important for samples with wide molecular weight distributions; the errors in the molecular weight averages vary from about 2 to 10%. Although a correction can be made for this effect, it is not a simple matter. Most published methods require considerable computation and good instrumental repeatability, and are not satisfactory for all types of samples. A review of methods of making the correction will be found in the paper by Duerksen[28].

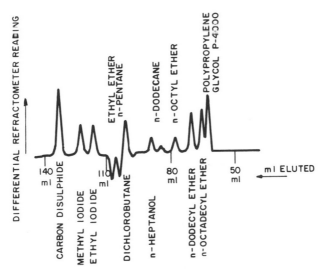

Fig. 5.14. Separation efficiency in a 12-ft column packed with 40 Å styrene–divinyl benzene gel[26]. (Reproduced by courtesy of the American Chemical Society.)

(2) Low-Molecular-Weight Compounds

Although the main application of exclusion chromatography has been in the separation of polymers, it has also been used extensively to separate low-molecular-weight compounds. Better separations can often be obtained by gas chromatography, but exclusion chromatography has two advantages; the separation can be made at room temperature and there is no lower or upper limit to the molecular weight of the solute. Figure 5.14 shows the separation by Hendrickson[26] of 13 compounds on a 12-ft column packed with 40 Å polystyrene–divinyl benzene gel.

Peaks due to impurities in the solvent are sometimes found in the low-molecular-weight region. When tetrahydrofuran is used as solvent, peaks due to oxygen, nitrogen, and water are often seen[29] (Fig. 5.15). This is due to the fact that the air and water contents of the injected sample solution differ from that of the main bulk of the tetrahydrofuran.

For low-molecular-weight substances the elution volume is a function of molar volume, plots of the logarithm of molar volume against elution volume being linear. Some polar solutes in polar solvents emerge earlier than would be expected from their molar volumes. This seems to be due to association of the solute and solvent molecules; alcohols and acids show this effect in tetrahydrofuran, but in a less polar solvent such as *O*-dichlorobenzene they behave normally.

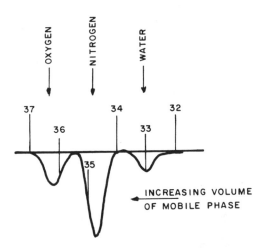

Fig. 5.15. Peaks due to impurities in mobile phase (tetrahydrofuran) using differential refractometer detector[29]. (Reproduced by courtesy of John Wiley and Sons, Inc.)

REFERENCES

1. Giddings, J. C., *Anal. Chem.* **39,** 1027 (1967).
2. Moore, J. C., *J. Polymer Sci.* **A-22,** 835 (1964).
3. Yau, W. W., Malone, C. P., and Suchan, H. L., Preprints, Division of Petroleum Chemistry, Am. Chem. Soc., **15,** A63 (1970).
4. Ginsberg, B. Z., and Cohen, D., *Trans. Faraday Soc.,* **60,** 185 (1964).
5. Porath, J., *Pure Appl. Chem.* **6,** 233 (1963).
6. Laurent, T. C., and Killander, J., *J. Chromatog.* **14,** 317 (1964).
7. Ogston, A. G., *Trans. Faraday Soc.* **54,** 1754 (1958).
8. Moore, J. C., and Arrington, M. C., Preprints, Third International Seminar on Gel Permeation Chromatography, Geneva (1966).
9. De Vries, A. J., Le Page, M., Beau, R., and Guillemin, C. L., *Anal. Chem.* **39,** 935 (1968).
10. Casassa, E. F., *Polymer Letters* **5,** 773 (1967).
11. Giddings J. C., Kucera, E., Russell, C. P., and Myers, M. N., J. *Phys. Chem. 12,* 4391 (1968).
12. Casassa, E. F., and Tagami, Y., *Macromolecules* **2,** 14 (1969).
13. Kwok, J., and Snyder, L. R., *Anal. Chem.* **40,** 118 (1968).
14. Haller, W. J., *J. Chem. Phys.* **42,** 686 (1965).
15. Haller, W. J., *Nature* **206,** 693 (1965).
16. De Vries, A. J., Le Page, M., and Beau, R., J. *Polymer Sci. C* **21,** 119 (1968).
17. Kelley, R. N., and Billmeyer, Jr., F. W., Preprints, Division of Petroleum Chemistry Am. Chem. Soc. **15,** A157 (1970).
18. Cooper, A. R., Cain, J. H., Barrall, E. M., and Johnson, J. F., Preprints, Division of Petroleum Chemistry, Am. Chem. Soc. **15,** A95 (1970).
19. Heitz, W., Klatyk, K., Kraffczyk, F., Pfitzner, K., and Randau, D., in *Proc. Seventh International Seminar on GPC, Monte Carlo,* p. 214 (1969).
20. Mair, B. J., Hwang, P. T. R., and Ruberto, R. G., *Anal. Chem.* **39,** 838 (1967).
21. Green, J. H. S., and Vaughan, M. F., *Chem. Ind.* 829 (1958).
22. Meyerhoff, G., *J. Polymer Sci. C* **21,** 31 (1958).
23. Brandrup, J., and Immergut, E. H., (eds.), *Polymer Handbook,* Interscience Publishers, New York (1966).
24. Benoit, H., Rempp, P., and Grubisic, Z., *Polymer Letters* **5,** 753 (1967).
25. Coll, H., and Gilding, D. K., *J. Polymer Sci. A-2* **8,** 89 (1970).
26. Hendrickson, J. G., *Anal. Chem.* **40,** 49 (1968).
27. Harmon, D. J., *J. Polymer Sci. C,* **8,** 243 (1965).
28. Duerksen, J. H., Preprints, Division of Petroleum Chemistry, Am. Chem. Soc. **15,** A47 (1970).
29. Alliet, D., *J. Polymer Sci. A-1* **5,** 1783 (1967).

FURTHER READING

Determann, H., *Gel Chromatography,* Springer-Verlag, Berlin (1968). (Deals with the theory and practice of exclusion chromatography in both aqueous and nonaqueous solutions.)
Cazes, J., *J. Chem. Ed.* **47,** A461, A505 (1970). (A good account of the practice of exclusion chromatography in nonaqueous solutions.)
Altgelt, K. H., *Advan. Chromatog.* **7,** 3–46 (1968). (A review of the theory of exclusion chromatography, now rather dated.)
Peaker, F. W., in *Chromatography,* Browning, D. R., ed., McGraw-Hill, London (1969). (A concise article on exclusion chromatography in nonaqueous solutions.)

Chapter 6

The Technique of Thin-Layer Chromatography

6.1. INTRODUCTION

The discovery of thin-layer chromatography (TLC) is attributed to Izmailov and Shraiber[1], who, in 1938, separated plant extracts by spotting them onto adsorbent layers produced by coating microscope slides with a slurry of the adsorbent with water and allowing to dry. Separation of the mixture into concentric zones was achieved by applying drops of certain solvents to the center of the spot.

In 1944, Consden *et al.*[2] introduced the method of paper chromatography and during the next twelve years over ten thousand papers involving this technique were published. Although based on the partition chromatography of hydrophilic solutes between an aqueous mobile phase and water held by the cellulose support, the techniques and apparatus developed form the basis of present-day thin-layer chromatography. Attempts to extend the technique to the chromatography of lipophilic solutes, e.g., by impregnating glass-fiber paper with adsorbents such as silica gel and alumina or coating strips of glass occurred during the early 1950's, notably by Kirchner and his co-workers[3-6] but it was not until 1956 that Stahl[7,8] recognized that standardized procedures, adsorbents, and equipment were needed to make the method generally applicable. As a result of his efforts, standard equipment* consisting of a coating apparatus, glass carrier plates, developing tanks, and adsorbents† became commercially available in 1958. A recent history of TLC is given in Ref. 9, which includes English translations of the original papers by Izmailov and Shraiber[1] and Stahl[8].

*C. Desaga, GmbH, Heidelberg, Germany.
†E. Merck, A.G., Darmstadt, Germany.

Since 1960 the number of publications describing the use of TLC has risen to an average of over 1000 per year, covering particularly the fields of biochemistry, pharmaceuticals, synthetic organic chemistry, and industrial organic analysis. Many reviews and several monographs have been written upon the subject, including two comprehensive works[57, 58] which cover most of the work published before 1966. Since that time the advances that have taken place have been mainly concerned with improvements in reproducibility and quantitative analysis. By 1964 TLC had achieved official status by its inclusion in standard methods in several pharmacopoeia[8].

Since we are dealing almost entirely with lipophilic solutes in non-aqueous solution, the techniques of paper chromatography—or its modern counterpart, using cellulose-coated glass plates—partition chromatography (normal or reversed phase), ion-exchange chromatography, exclusion chromatography, or the separation of inorganic ions will not be mentioned any further. Suffice it to say that the apparatus and techniques presently described are generally applicable in all thin-layer techniques. Only the separation mechanisms differ. It is certainly not the intention to cover in this chapter that which has been adequately dealt with elsewhere (see the list of further reading). Instead, an account will be given of the technique of thin-layer adsorption chromatography as it is carried out in the authors' laboratory, based on extensive experience since 1961, together with a discussion of some recent developments.

6.2. OUTLINE OF TECHNIQUE

TLC is a simple, rapid separation technique using inexpensive, portable apparatus, and is therefore particularly attractive to those laboratories that operate on low capital budgets, and is also very amenable to field use. It is the technique to be used where a large number of samples are to be examined and where the components of interest are easily separated from each other.

The efficiency of a conventional chromatographic separation on thin layers is seldom greater than 1000 theoretical plates and although separations can be improved either by modification of the stationary phase to enhance selectivity, by modification of the mobile phase, e.g., gradient elution, or by multiple or stepwise development (unidirectional or bidirectional), or by lowering the temperature, the basic simplicity of the technique is reduced and analysis times are increased. Complex separations are therefore preferably carried out in high-efficiency columns, where control over chromatographic conditions is more easily achieved.

In TLC, 2–10-μl portions of the sample components are applied in the form of 0.1–1% solutions near the bottom of a plate (glass, metal, or plastic) coated with a thin layer, usually of 100–300 μm thickness, of stationary

phase. The plate is placed in a chamber containing a depth of a few millimeters of the appropriate mobile phase, which travels through the layer by
capillary action. In doing so, sample components travel at different speeds
through the layer depending on their adsorption coefficients. Development
is stopped by removing the plate from the chamber and evaporating the
mobile phase. The position of the separated solutes is located, if not already
visible when viewed either by normal or ultraviolet light, by spraying with a
chromogenic reagent. The position of the solute spot is described by measuring its R_F value, defined as

$$R_F = \frac{\text{Distance moved from point of application by solute}}{\text{Distance moved from point of application by mobile phase}}$$

The R_F value of all adsorbed solutes is therefore less than 1.00. For
convenience, hR_F values are sometimes quoted, hR_F being defined as $100R_F$.

The R_F value of a solute is related to its adsorption coefficient K by the
expression

$$R_F = \frac{1}{1 + (W_a/V_s)K}$$

where W_a and V_s refer to the amount of stationary and mobile phases,
respectively. The derivation of this expression assumes that the ratio W_a:
V_s is constant across the whole of the developed plate, that the mobile phase
has a uniform composition, and that the mobile phase moves solely by
capillary migration, i.e., adsorption of solvent vapor by the stationary phase
is avoided. These conditions are best achieved in practice by development in
sandwich chambers and allowing the mobile phase to overrun (see later).
If overrunning is not allowed to occur, then the R_F value is better described
by the expression

$$1.1R_F = \frac{1}{1 + (W_a/V_s)K}$$

An alternative method for measuring the position of a spot is to relate
it to the position of a reference compound X chromatographed simultaneously. Its position is then given by the R_X value defined as

$$R_X = \frac{\text{Distance moved from point of application by solute}}{\text{Distance moved from point of application by compound } X}$$

In contrast to R_F values, R_X values of greater than 1 are possible.

Tables of R_F values collected from the literature are published regularly
in the *Journal of Chromatography*.

Quantitative analysis in TLC is carried out by chromatographing
standard solutions of the component under investigation alongside the test

sample. The amount of material in a spot is related to the area and intensity of the revealed spot, and standards and samples can either be compared visually or measured instrumentally, e.g., by direct spectrophotometry, densitometry, etc.

The material in a spot may be eluted from the stationary phase for further examination. This could either be for further quantitative measurements by spectroscopic methods, for the identification of an unknown species by infrared or mass spectroscopy, or for further separation, e.g., by gas chromatography. For these purposes the sample size can be increased by using thicker layers (up to 2 mm thickness), applying the sample as a band, and using wider plates (up to 1 m).

6.3. THE STATIONARY PHASE

In principle, all stationary phases that have been used in column chromatography can be used in TLC. Although adsorbents, in particular silica gel and to a lesser extent alumina, are the most widely used stationary phases, numerous examples have been reported using normal partition (cellulose), reversed phase partition (e.g., kieselguhr impregnated with high-boiling, nonpolar organic liquids), ion-exchange, and exclusion stationary phases.

Adsorbents for TLC are readily available commercially [see Stahl[57]]. The properties of these adsorbents have also been discussed in Chapter 3. As was mentioned in this chapter, the properties of the same adsorbents differ widely from manufacturer to manufacturer. The chromatographer is therefore advised to standardize on one particular source and get to know the properties of his adsorbent. This is particularly important where TLC is to be used in conjunction with column chromatography.

Commercial adsorbents for TLC have particle diameters in the range 1–40 μm. They may contain additives, e.g., gypsum (5–20%) or starch (2–5%) as binders, which aid in the production of a uniformly coated plate, or fluorescent indicators, such as zinc silicate, which aid in the detection of fluorescent quenching materials when viewed under ultraviolet light. The surface properties may be further modified to confer specific adsorption characteristics (see Chapter 3).

Many commercial adsorbents contain inorganic impurities such as iron as well as solvent-extractable organic impurities which interfere in the identification of solutes eluted from TLC plates. We have found that Silica Gel H, HF254, or HR (Merck) to be satisfactory general-purpose adsorbents useful for both TLC and high-efficiency column work. These adsorbents contain no added binder, are free from solvent-extractable organic matter, and produce uniformly coated TLC plates which can be satisfactorily used with most organic solvents. Silica Gel HF254 contains an inorganic fluorescent indicator which exhibits a green fluorescence when viewed under short-wave (254-nm) ultraviolet light. This enables solutes that absorb the fluorescence at this wavelength to be detected as dull mauve spots on a green fluorescent background. Silica Gel HR is a highly purified adsorbent and is especially suitable for use where the separated solutes are to be eluted from the adsorbent for further identification.

Although adsorbents containing binders produce firmer layers, the adsorption properties are modified, development times are increased, and in the case of gypsum, plates must be coated within 4 min of preparing the slurry before dehydration of the gypsum occurs.

6.4. PREPARATION OF TLC PLATES

6.4.1. Choice of Support

TLC plates consist of thin layers of adsorbent (0.1–10 mm thickness) spread on a supporting base made of glass, aluminum foil, or plastic.

Adsorbent-impregnated glass fibers are also commercially available, but are rather fragile.

For preparing one's own layers, glass plates of 2–4 mm thickness are invariably used. Glass has the advantage that it can be easily cleaned, repeatedly used, and withstands the usual solvents and revealing reagents, including most corrosive substances. Three standard sizes are available: 20 × 20 cm, 10 × 20 cm, and 5 × 20 cm, although plates up to 1 m wide are also available. Borosilicate glass can be used if the developed chromatogram is to be subjected to high temperatures ($>150\,°C$), e.g., in using charring techniques to reveal the chromatogram. After use, plates are cleaned by washing under running water, immersed for 24 hr in a tank of distilled water containing 1 % of an aqueous surfactant detergent, and finally washed in distilled water and left to dry.

6.4.2. Preparation of Slurry

The adsorbent is applied to the plate as an aqueous slurry. To coat five 20 × 20 cm plates or twenty 20 × 5 cm plates with silica gel to a wet layer thickness of 0. 25mm, 27.5 g of Silica Gel H, HF254, or HR are shaken vigorously for 1 min in a stoppered 250-ml conical flask with 60 ml of water. To coat the same number of plates to a wet layer thickness of 0.75 mm, the corresponding quantities are 80 g of adsorbent to 180 ml of water. The slurry may be stored indefinitely in a well-stoppered flask. When using adsorbents from other sources the manufacturer's recommendations are followed. Adsorbents containing gypsum as binder must be spread within 4 min from the addition of the water, since after this period has elapsed the gypsum begins to "set."

6.4.3. Coating the Plate

There are four basic ways of coating a plate: (a) by pouring the requisite amount of slurry onto the plate and attempting to obtain an even coating by appropriately tilting the plate to and fro, or by spreading with a ruler or glass rod; (b) by dipping the plate into the slurry, withdrawing the plate, and allowing to drip dry (see the section on choice of mobile phase); (c) by spraying the slurry onto the plate with an air gun; and (d) by using a mechanical spreading device.

The reproducibility of layer thickness obtained by the fourth method is much greater than that achievable by the first three. It is also by far the most convenient method when several plates are to be prepared. Sample applicators are readily available commercially. [Stahl[57] lists 156 manufacturers and suppliers of TLC equipment.]

Mechanical spreading devices are basically of two types:

(a) In the "Kirchner type" the plates are successively pushed under an immovable trough containing the slurry. The height of one edge of the trough can be adjusted above the plate so as to provide the desired wet layer thickness. The manual versions of this spreader are best operated by two people, but motorized versions are available. This is the most convenient type of spreader to use where the daily consumption of plates exceeds 20.

(b) In the "Stahl type" (Fig. 6.1) the trough containing the slurry is

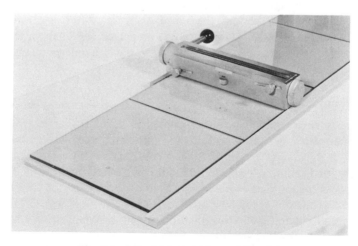

Fig. 6.1. Adjustable spreader and template.

Fig. 6.2. Drying rack and storage cabinet.

drawn smoothly over a series of plates aligned along a plastic template. The gap between the fourth side of the trough and the plates can be adjusted to produce coatings between 0–2 mm thickness as required.

This spreading device, which is simple, quick, and easy to use, has given us trouble-free operation for over ten years. As the spreader passes from one plate to the next there is a tendency to form a ridge due to slight variations in the thickness of the glass. By using the adsorbent to water ratio previously mentioned, which contains slightly more water than the manufacturer's recommendation, it has been found that the ridges can be easily dispersed by gently tapping the template on the bench immediately after the plates have been coated.

A modification of this type of applicator is available (Desaga) for the preparation of gradient layers of two stationary phases X and Y ranging in composition from pure stationary phase X at one end of the plate to pure Y at the other end. This enables one to choose an optimum composition of adsorbent for a given sample and mobile phase.

6.4.4. Drying and Storage of Plates

A freshly coated plate has a bright, reflecting surface due to the presence of surface water. Plates in this condition should be left undisturbed on the template until the surface water has evaporated and the silica gel has a dull, transparent appearance. This takes 15–30 min in the case of layers of 250 μm thickness, and correspondingly longer in the case of thicker layers.

The plates are then inserted into the metal drying rack (Fig. 6.2), which in turn is placed so that the plates are held vertically into a well-ventilated drying oven maintained at 105–110°C for 60 min. At the end of this period the plates are transferred to the wooden storage cabinet (Fig. 6.2), which contains self-indicating silica gel. The layer has a bright, translucent appearance.

A satisfactory plate will appear to be uniformly coated when viewed under transmitted light. Ridges and lumps of adsorbent should be absent. To avoid disturbing the layer by handling of the plate, it is advisable to remove a 4-mm margin of adsorbent around the plate using the thumb and forefinger.

6.4.5. Precoated Plates

In recent years an increasing number of precoated plates have appeared upon the market. A partial list of commercial sources is given in Table 6.1. Although we describe these as "plates," the support may take the form of plastic sheets, metal foil, or impregnated glass fiber as well as glass plates. Plates coated with silica may contain binder or fluorescent indicator. In addition to silica-coated plates, plates coated with alumina, cellulose, polyamide, and polycarbonate are also available.

In contrast to having to prepare one's own plates, perhaps photograph them, and then clean them afterwards, the opportunity to select a plate from a packet, to stick the developed and revealed chromatogram into a notebook, or else discard it has its obvious attractions. This has been described as "instant thin-layer chromatography." There are, nevertheless, a number of other advantages and disadvantages that should be taken into account when selecting a particular product. These are as follows.

(a) The present cost of a 20 × 20 cm precoated plate is about $0.50, compared with $0.05 for a hand-coated plate. This latter amount reflects the price of the adsorbent, but not the reusable glass support or spreading equipment. On this basis, the high cost of using a precoated plate is only justifiable where labor charges are high. The time taken to prepare five 20 × 20 plates, carry out the analysis, and suitably record the finished chromatogram will be about an hour.

(b) The chromatographic properties of the different types of plate differ from each other and from home-made plates. Table 6.1 compares the layer thickness (which determines the sample size), the nature of the binder, the speed of analysis as given by the time it takes for toluene to ascend a height of 10 cm when a plate is developed in a sandwich chamber, and the R_F values of the components of a test dye mixture when developed in toluene. It can be seen that wide variations occur.

(c) The handling properties of different types of precoated plates can vary widely. We have found that (1) silica gel-impregnated glass fiber sheets flaked so badly on handling that the great majority in a given packet were unusable; after development they lost their rigidity and behaved in much the same way as a piece of wet blotting paper; (2) aluminum-coated foils curled up irreversibly on gently heating in an oven and became unusable; (3) plastic-coated sheets possessed firm layers; those bound by polyvinyl alcohol were firmer than the starch-bound ones, and could be written upon by pencil; however, adsorbent coatings were thin (approximately 100 μm) and consequently their sample capacity was low; along with aluminum foils, they have the advantage that they can be easily cut to any desired shape with scissors and conveniently be preserved in laboratory notebooks; (4) the behavior of precoated glass plates depends upon the source; with the first batch of one company's plates we received, most of the adsorbent fell off before they could be used; a recent batch, however, was more satisfactory; some precoated silica gel plates, on the other hand, possess a very uniform layer firmly bound to the glass support and can withstand excessive handling; in fact, the adsorbent can only be removed by chipping off with the sharp edge of a spatula and can be written on by a pencil or ball point pen; the products of E. Merck A.G. are particularly recommended.

(d) Some revealing reagents cannot be used with certain types of precoated plate because of their reaction with the support or the binder. Plastic sheets cannot be heated above about 130°C, and so revealing chromatograms by charring, for example, is ruled out.

(e) The choice of a precoated plate for the separation and isolation of solutes requiring spectroscopic measurements is restricted because of the high risk of contamination by foreign organic matter. This eliminates plastic sheets and all adsorbents containing organic binders. Only absolutely pure adsorbents can be used.

6.4.6. Conclusions

We will summarize this section by stating, with reasons, our current preferred choice of silica-coated plates. For reasons of cost and convenience our first choice would be a precoated plate.

(a) For general qualitative work, Merck Silica Gel H or HF 254 (without and with fluorescent indicator, respectively) coated glass plates are used. Development times are short and most revealing reagents including corrosive ones can be used. A disadvantage is that the adsorbent can easily be rubbed off and the chromatograms cannot conveniently be preserved.

(b) For quantitative work, Merck Silica Gel F or F254 (without and with fluorescent indicator, respectively) precoated glass plates are used, because their strongly bound layers can easily withstand the extra handling involved. For the same reason they are very useful for field work and chromatograms can be stored indefinitely in the dark. Because of the very small particle size of the adsorbent, flow rates are low, but sharp spots, ideal for direct quantitative analysis on the layer, are obtained. Most corrosive revealing reagents can be used, although above 130 °C reactions with the organic binder can occur. The organic binder also interferes with spectroscopic measurements on the spots after elution from the adsorbent.

(c) To identify solutes eluted from a TLC plate using spectroscopic methods, hand-coated glass plates of Merck Silica Gel HF254 or HR are used since these adsorbents have been shown to be relatively free from organic impurities.

6.5. APPLICATION OF THE SAMPLE

Samples are applied in the form of 0.1–1 % solutions in a nonpolar, volatile solvent. The solvent should be nonpolar to minimize the spreading at the point of application. It should also be low boiling (i.e., boiling point less than 120 °C), so that it quickly evaporates before chromatography commences. The solvent must be sufficiently strong so that the solute does not crystallize out of solution before being adsorbed on the adsorbent.

6.5.1. Analytical Scale

The technique of applying the sample is illustrated in Fig. 6.3. An activated plate is taken from the storage cabinet and stored in a chamber containing saturated aqueous sodium bromide solution for 15 hr to bring the plate to constant activity. It is then placed on a flat bench and partly covered with a plastic template so as to reduce the risk of further deactivation by atmospheric moisture. The template is so positioned as to leave the bottom 2 cm exposed to allow application of the sample. Then 2–10 μl of the sample solutions are applied from a 10-μl Hamilton-type microsyringe as a series of spots 1.5 cm above the bottom edge of the plate. Care must be taken not to disturb the adsorbent surface when applying the solution. This is no problem with the Silica Gel F type of precoated plates. The diameters of the spots should be between 2 and 4 mm, and the spot centers 10–15 mm apart. The template will assist in spacing the spots appropriately. When spotting has been completed the limit to which the mobile phase is to run is marked out by scouring the adsorbent with a sharp point, completely removing the adsorbent so that solvent cannot flow past this line. The distance between the point of application and the solvent front is usually standardized at 10 cm. This distance represents an optimum for good resolution and short analysis time.

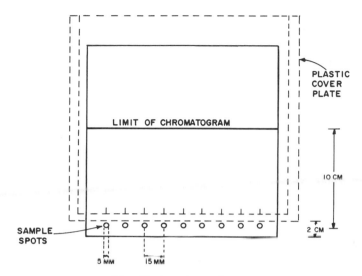

Fig. 6.3. Spotting a plate.

For quantitative work sample and reference solutions are spotted alternately. The spot sizes must be identical and the compositions of the sample and reference solutions should match each other as closely as possible. In particular, the solvent compositions must be the same. It has been shown[10] that a major source of error in quantitative analysis in TLC lies in the repeatability of applying a constant volume of solution of the layer, standard deviations of up to 10% at the 1-μl level having been reported. These large errors have been attributed to solvent creeping back up the outside of the needle, particularly noticeable with polar solvents such as methanol, and to capillation, which occurs when the tip of the needle is allowed to touch the surface of the adsorbent in an effort to get rid of an adhering drop. This can cause a further amount of solution to be drawn from the lumen of the needle. In an effort to reduce these errors, Bridget and Relph designed a machine, the Chromaplot (Burkhard Manufacturing Co., Ltd., Rickmansworth, Hertfordshire., England) in which small, repeatable volumes of solution are ejected from the needle onto the plate. Using this device standard deviations were reduced to less than 3%[10].

For densitometric work (see the section on quantitative analysis) we have found that application of 5 μl of solution from a 10-μl Hamilton microsyringe, the needle of which has been sawn off square to leave a 2 cm length, gives a similar standard deviation of 3%. In general, we apply 5 μl of 0.1, 0.2, 0.4, 0.6, 0.8, and 1.0% solutions of the reference compound (i.e., 5, 10, 20, 30, 40, and 50 μg solute) and for greatest accuracy dilute the sample solutions so that the unknown concentration to be measured lies toward the lower end of this range.

6.5.2. Preparative Scale

The separation of sample sizes greater than 50 μg requires, in the first instance, that the sample solution be applied as a narrow band, and ultimately the use of thicker layers of adsorbent. Up to 50 mg of sample can be handled on a 20 × 20 cm plate coated with a layer thickness of 750 μm. A further increase in the layer thickness results in a serious loss of resolution, while a further increase in sample size causes tailing because the linear capacity of the adsorbent is exceeded. The separation of sample sizes greater than 50 mg should be attempted either on a number of plates developed in parallel or by column chromatography.

Application of a large sample by repetitive spotting is both tedious and unsatisfactory since, even with the steadiest eye and firmest hand, an irregular band results. A convenient method is to use a commercial band applicator. Although a number of such devices are available, it can be said that the more expensive, automated machines offer no advantages over simpler, hand-operated devices at one-tenth of the cost. An example of the latter type is shown in Fig. 6.4. A micrometer-screw-driven syringe mounted on an adjustable length of carriageway delivers 2–20 μl cm^{-1} (depending on the syringe capacity) via a length of capillary tubing shaped to rest lightly on the adsorbent layer. The tube is lifted clear of the adsorbent surface on the return stroke. The sample is usually applied as a 5–10% solution in a nonpolar solvent. This means that 50 mg of sample can be applied to a 20 × 20 cm

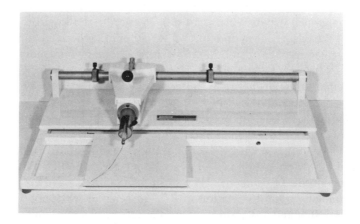

Fig. 6.4. Streak applicator.

Fig. 6.5. Development chambers.

plate in only two strokes. The solvent must be allowed to evaporate between the application of successive strokes.

6.6. CHOICE OF MOBILE PHASE

The factors involved in the selection of a suitable mobile phase have been discussed in Chapter 3, and Table 3.4 lists solvents in order of increasing eluent strength.

In practice, the preferred method for choosing a suitable solvent for both TLC and column work is based on the work of Peifer[11]. A pair of microscope slides back-to-back is dipped into a slurry of adsorbent, e.g., 35 g of Silica Gel H in chloroform–methanol (65–35 vol.) which is stored in a tall glass vessel. The slides are withdrawn, separated, and allowed to dry quickly. Then 2 μl of a 1 % solution of the sample under test is applied to the plate and developed in a reagent bottle containing a small amount of solvent of low eluent strength. This usually takes only 2–3 min, after which time the plates are taken out and quickly dried and the spots located with a suitable revealing reagent. This is repeated using solvents of increasing strength, using binary or ternary mixtures if necessary, until the R_F values of the components of interest lie between 0.3 and 0.8. The opportunity is also taken to try out possible revealing reagents.

6.7. DEVELOPMENT OF CHROMATOGRAM

The development of the chromatogram is normally carried out by ascending chromatography, in which the mobile phase rises through the layer

Fig. 6.6. Sandwich chamber.

by capillary action. Two types of development chamber are used, the tank (or jar, for smaller plates) (see Fig. 6.5) and the sandwich chamber, in which a cover plate is combined with the TLC plate to form, in effect, a rectangular tank (Fig. 6.6).

6.7.1. Tank Development

Mobile phase is poured into the tank to a depth of 5 mm. A 15 × 40 cm sheet of resin and fat-free filter paper is laid in the form of a square-shaped U so as to line the two long sides and bottom of the tank, and thoroughly wetted with the solventy by shaking. This assists in saturating the atmosphere of the tank with solvent vapor. The tank should be left to stand for 30 min at a constant temperature, away from direct sunlight, to allow the atmosphere in it to come to equilibrium. The effect of having an unequilibrated tank means that during chromatography, evaporation of mobile phase occurs from the face of the plate, causing more mobile phase to flow through the adsorbent before the predetermined limit of the solvent front has been reached. R_F values in an unsaturated atmosphere are therefore higher and less reproducible than in a saturated atmosphere. Furthermore, since evaporation of solvent proceeds more rapidly at the edge of the plate than in the center, a concave solvent front is formed. Consequently, the R_F value of a given solute is greater at the edges of the plate than in the center (Fig. 6.7).

It should be noted that in the case of a multicomponent mobile phase the composition of the vapor will differ from that of the liquid. Until equilibrium has been established, therefore, the composition of the mobile phase may be continually changing, especially where one component is much more volatile than the others. If this component is also polar, it will be readsorbed on the unwetted portion of the plate and modify the adsorbent properties. This effect has been exploited by de Zeeuw[12] in "vapor programmed" TLC.

A typical tank for accommodating 20 × 20 cm plates is made of glass, inner dimensions 21 × 21 × 9 cm, with a flat internal base and a closely fitting ground glass lid. For accommodating 5 × 20 cm plates a cylindrical glass vessel 8 cm in diameter and 22 cm in height is suitable. Two plates can be developed simultaneously by leaning one plate against each side wall. Care must be taken to see they are not in contact with each other. On dipping the plates into the mobile phase, it is important to see that the spot at the starting point is above the level of the solvent in the tank; otherwise the sample will diffuse away.

6.7.2. Sandwich Chambers

Problems associated with solvent vapor are much reduced by using a sandwich chamber, and development chambers of this type are preferred for accurate work, i.e., quantitative analysis and measurement of R_F values.

In the Desaga apparatus designed by Stahl, a 0.5-cm strip of adsorbent is removed all around the edges of a 20 × 20 cm plate, which forms one wall of the chamber. A cover plate is carefully placed on top of the TLC plate to form the opposite wall. The side walls and lid 2–3 mm thick are three narrow glass strips fused onto the cover plate. The sandwich thus formed is held together with two strong clips and the bottom is dipped into a trough con-

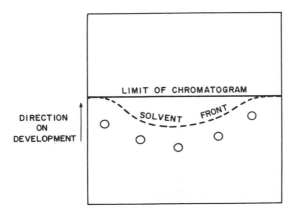

Fig. 6.7. Illustration of the "edge effect" due to development in an unsaturated atmosphere.

taining 25 ml of mobile phase. Development must be carried out at constant temperature, otherwise mobile phase may evaporate from the face of the plate and condense on the cover plate. Condensation may also occur if the mobile phase has a high heat of adsorption. Either situation leads to poor reproducibility.

Sandwich chambers are now available from several manufacturers, some models having been designed especially for use with precoated plates.

6.7.3. Other Methods of Development

(i) Descending Development. Mobile phase is fed, by means of a wick dipping into a solvent reservoir, to the top edge of a vertical TLC plate. This technique shows no advantages as regards development time or resolution over ascending development. Since the apparatus is more cumbersome, this method is rarely used.

(ii) Circular Development. Mobile phase is slowly applied to the center of a circular sample spot in the middle of a horizontal TLC plate. Sample components move outward in the form of concentric rings. It has been used with microscope slides for quickly selecting a suitable mobile phase[8], but otherwise the chromatograms are difficult to evaluate. The solvent migration rate can be accelarated by applying a centrifugal force, i.e., by spinning the plate in a horizontal plane[13].

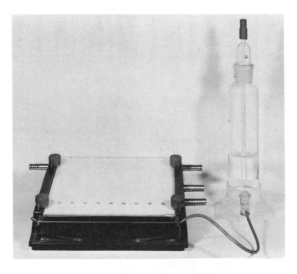

Fig. 6.8. Horizontal BN chamber.

(iii) Horizontal Development. As a result of work by Brenner and Niederwieser[14], the apparatus known as the "Horizontal BN-Chamber" is commercially available (Desaga) and is shown in Fig. 6.8. Solvent from the Drechsel bottle enters the trough on the left through a Teflon tube and then reaches the TLC plate via a paper tongue. The plate rests on a metal block through which cold water may be circulated. The plate is covered by an 18-cm-long cover plate to prevent evaporation of mobile phase from the layer.

(iv) Continuous Development. This, in effect, is a method of lengthening the plate to improve the resolution of slowly moving solutes. It can conveniently be carried out using the BN-Chamber described above. The top 2 cm of the TLC plate juts out beyond the cover plate and rests on a metal block which can be heated by circulating hot water or steam or by inserting an electrically heated filament. In this way, as the solvent emerges from beneath the cover plate it is evaporated. Since there is no recognizable solvent front in this technique, a reference compound X must be chromatographed simultaneously and R_X values measured (see the section on outline of technique).

(v) Multiple Development. In this technique the plate is developed a number of times in the same solvent, the plate being dried between each development. It is of value in improving the separation of slow-moving, poorly resolved solutes. The final R_F value ^nR_F can be predicted from the R_F after one development, ^1R_F , by the expression

$$^nR_F = 1 - (1 - ^1R_F)^n$$

where n is the number of developments.

This technique has been studied by Halpaap[15] in conjunction with TLC on the 10-g scale.

(vi) Stepwise Development. This technique is used to separate a mixture containing solutes of widely different polarities. For example, to separate polar additives in a lubricating oil[16] the plate is first developed with a nonpolar mobile phase such as toluene for the full length of the plate. This moves the hydrocarbon base oil to the top part of the plate, leaving the polar additives undisturbed on the start point. The plate is then removed from the developing chamber, the mobile phase evaporated, and the plate redeveloped the normal distance of 10 cm in a polar mobile phase to chromatograph the polar additive mixture.

(vii) Gradient Elution. In this technique the composition of the mobile phase is changed continuously during the course of development. Wieland and Determan[17] have developed a chamber (Fig. 6.9) in which the polar mobile phase to be added is pumped in through the capillary tube on the left-hand side of a circular chamber and is mixed with the nonpolar mobile

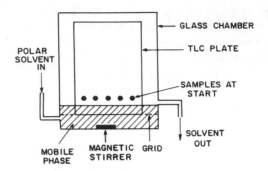

Fig. 6.9. Apparatus for gradient elution TLC
[after Wieland and Determan[17]].

phase, already in the chamber, with the help of the magnetic stirrer. The TLC
plate stands on the grid as shown. The volume of the mobile phase is kept
constant by incorporating an overflow.

(viii) Two-Dimensional Development. This technique is used for the
examination of complex mixtures. The sample is applied in one corner of
a 20 × 20 cm plate and developed the full length of the plate in the normal
manner, the object being to space out the components as much as possible.
The plate is then turned through 90° and developed again in the second direction. To give the full advantage of this technique, the chromatographic
conditions should be changed for the second development. This can be
effected in one of two ways: (1) by altering the nature of the stationary
phase; one can either completely remove the unused adsorbent and recoat
with a different adsorbent, impregnate the unused adsorbent with a high-
boiling liquid so as to make use of the partition mechanism; or employ thin-
layer electrophoresis; (2) by chemically altering the nature of the partially
separated solutes before rechromatographing in the second direction using
the same mobile and stationary phases as in the first development; unchanged
solutes will be on a line drawn diagonally across the plate, but solutes that
have undergone a reaction will have different R_F values when developed
in the second direction. The chemical reaction may be effected by radiation,
heat, reactive gases (e.g., bromination), or by treatment with a reactive chemical reagent.

(ix) Polyzonal Development. This occurs when the mobile phase consists of two or more solvents of widely differing polarities and the phenomenon of solvent demixing is observed (see Chapter 3). This is because the
adsorbent has a stronger affinity for the more polar solvents, with the result
that secondary solvent fronts are formed. The mobile phase above the secondary solvent front is richer in the less polar components.

Niederwieser and Brenner[18] have exploited this situation as an aid in choosing an optimum solvent system for a given separation. The sample is spotted diagonally across a 20 × 20 cm plate. If the plate is developed in a multicomponent mobile phase in which solvent demixing occurs, each sample spot is treated in turn to a series of solvent fronts of differing composition. By studying the behavior of the sample in each zone of mobile phase, the correct choice of mobile phase can be made to effect an optimum separation of the sample mixture.

Recent work[19] on the influence of solvent vapor in TLC has shown that if development of the plate is carried out in unsaturated chambers, then separations with multicomponent solvents are more effective than in saturated chambers. The improvements are caused by a concentration gradient mainly dependent on the rate of evaporation of the solvent components and on their affinity for the adsorbent used.

A vapor-programming chamber has been developed[12], in which the plate is placed, adsorbent facing downward, over a series of troughs containing solvents whose vapor compositions can be controlled so that the desired polarity gradient can be obtained along the length of the plate. The chromatogram is developed horizontally and rigorous temperature control must be maintained if repeatable results are to be achieved.

The theory and technique of gradients in TLC (adsorbent, solvent, and solvent–vapor gradients) have recently been reviewed[20–22]. Both the theory and practical operations are complex, and detract from the basic and attractive simplicity of TLC. Separations requiring this degree of resolution can be more quickly and simply carried out by gradient elution high-efficiency column chromatography, where chromatographic conditions can be more readily controlled.

6.8. REPRODUCIBILITY OF R_F VALUES

6.8.1. Introduction

Tables of R_F values are regularly published (e.g., in the *Journal af Chromatography*), which, taken in conjunction with behavior with respect to specific revealing reagents, are used to make a tentative identification of a separated solute. The reproducibility of R_F values is much poorer at present in TLC than with paper chromatography, but as progress is made in the standardization of the technique[23], R_F values reproducible to within 0.05 should be readily achievable. The factors affecting R_F values (other than solute structure) have been studied by many workers and was the subject of a recent symposium (see the list of further reading). A list of the known factors that could possibly affect the reproducibility of R_F values, and the precautions that should be taken, are now given.

Table 6.1. Comparison of Precoated Plates with a Home-Made Silica Gel H Plate

Manufacturer	Trade name	Support	Film thickness, μm	Binder	Development time,[a] min	R_F values[b]		
						Yellow	Red	Blue
E. Merck A.G. Darmstadt, Germany	Silica Gel H	Glass[c]	250	None	15	0.48	0.17	0.03
	Silica Gel F	Glass	250	Organic	37	0.54	0.18	0.08
	Type 5553	Aluminum	250	Organic	38	0.54	0.18	0.08
Distillation Products Industries Eastman Kodak Co. Rochester, N.Y.	Chromagram	Plastic	100	Polyvinyl alcohol	45	0.76	0.34	0.23
Machery-Nagel & Co. Düren, Germany	Polygram	Plastic	100 or 200	Starch	26	0.56	0.19	0.08
Mallinckrodt Chemical Works. St. Louis, Mo.	Chromar	Glass fiber	500 or 1000	None				
Gelman Instrument Co. Ann Arbor, Mich.	ITLC-SA	Glass fiber		None	17	0.88	0.52	0.40
Camag A.G. Muttenz, Switzerland	DSF-A	Glass	250	Not known	21	0.47	0.15	0.08

[a] Time taken for toluene to ascend 10 cm in S-type chamber.
[b] Test dye mixture developed with toluene in S-type chamber.
[c] Home-made plate, for comparison.

6.8.2. Factors Affecting Reproducibility

(i) Quality of Adsorbent. Probably Stahl's greatest contribution to TLC was his insistence on the standardization of adsorbents, and nowadays many TLC adsorbents bear the label "according to Stahl." Nonetheless, products from different manufacturers can still be expected to show variations due to difference in surface properties, surface area, and particle size. This was illustrated in Table 6.1. The situation is further complicated by the recent flood of precoated plates on the market. Satisfactory reproducibility will only be obtained when TLC adsorbents have been fully standardized and, preferably, when all binders have been eliminated.

(ii) Activity of Adsorbent. Having chosen a standard adsorbent, it is essential to control its activity. The R_F value of a solute developed in a given solvent can range from 0.00 on a fully activated plate to 1.00 on a plate equilibrated in a chamber in which the relative humidity is 90%. A plate left to dry in the atmosphere after coating has an unknown activity, dependent upon the relative humidity of the atmosphere. On the other hand, Dallas[24] found a plate that had been activated by heating in an oven at 110°C took up more than half the total amount of moisture adsorbed at equilibrium (in an atmosphere of about 50% relative humidity) in about 3 min, and that even breathing on a plate during the spotting process could markedly affect R_F values. He therefore recommends equilibrating the plate by allowing it to stand for 15 hr in a closed chamber containing a saturated aqueous solution of sodium bromide, corresponding to 58% relative humidity at 20°C. Spotting can safely be carried out without any significant change in activity occurring, providing the plate is covered by a template. Immediately after samples have been applied the TLC plate must be quickly made up into a sandwich chamber and the chromatogram developed.

(iii) Layer Thickness. Spreading devices deliver a constant wet-layer thickness, usually standardized at 250 μm, but the depth of the dried layer depends upon the speed of spreading, the consistency of the slurry, the water content and particle size of the dried layer, and the evenness with which the layer has been spread over the plates. Although some workers claim to have detected a small increase in R_F value with increase in layer thickness, it now seems that layers of constant activity developed in an S-chamber show no variation in R_F values with layer thickness within the range 0.1–3 mm.

(iv) Purity of the Mobile Phase. Solvents of the highest purity must be used for the measurement of accurate R_F values. The presence of a small amount of impurity of different polarity can have a big effect on the R_F value. Purity checks should also be taken even when switching to a different batch of solvent from the same manufacturer. When using mixed solvents as the mobile phase a fresh mixture should be used for each chromatographic

run to allow for changes in composition either because of differential evaporation or adsorption occurring during each development, or because of chemical interaction between the solvent components.

(v) Effect of Solvent Vapor. This aspect has already been fully discussed under types of development chambers. To ensure rapid saturation of solvent vapor within the vicinity of the solvent front and to minimize modification of the adsorbent by solvent vapors, sandwich chambers must always be used for the accurate measurement of R_F values. Since with sandwich chambers solvent demixing is greatly enhanced, it is advantageous to use mobile phases consisting of a single, pure solvent. Even chloroform (which normally contains 1% ethanol as a stabilizer) will form two distinct solvent fronts in a sandwich chamber.

(vi) Effect of Temperature. An ambient temperature increase of 10 °C can cause a slight increase in R_F value because of increased evaporation of solvent from the face of the plate. Similar effects occur if the mobile phase has a high heat of adsorption. If the mobile phase is a mixed solvent, then a change in composition might occur due to differential evaporation. Variations in R_F values using an S-chamber are negligible if the temperature is controlled within 2 °C.

(vii) Depth of Mobile Phase. The distance between the point of application of the sample and the level of the mobile phase in the development chamber can affect the R_F value of a spot. This is of importance where the mobile phase consists of a mixture of solvents of different polarities and its eluent strength (i.e., composition in the adsorbent) depends upon the distance traveled. The mobile phase depth should therefore be standardized at 0.5 cm and the sample spotted 1.5 cm above the lower edge of the plate.

(viii) Nature of Solvent Front. As the mobile phase advances through the layer the ratio of mobile phase to stationary phase is not constant but is smaller near the solvent front than at some distance behind it. Dallas[24] found that if the plate is allowed to stand for 15 min after the mobile phase first reaches the layer limit, then the R_F values, although a little higher than if measured immediately, are more constant and reproducible. This is because the ratio of liquid to stationary phase has had time to become constant over the whole length of the chromatogram.

The R_F values measured in this way are independent of the distance between the point of sample application and the solvent front if the mobile phase is a single pure solvent. With a mixed solvent, however, there is often a solvent gradient up the plate and R_F values are no longer independent of the distance the mobile phase has traveled. This distance has therefore to be standardized, usually at 10 cm.

(ix) Mobile Phase Velocity. The R_F value of a particular solute in a given phase system will depend upon the mobile phase velocity because the

rate of attainment of equilibrium is not instantaneous, i.e., depends upon whether ascending, descending, or horizontal development is used. In the case of ascending development the angle the plate makes with the vertical could also affect the R_F value. By using the sandwich chamber in the vertical position, this problem does not arise. The mobile phase velocity is also affected by the particle and pore size of the adsorbent.

(x) Sample Size. The R_F value is independent of sample size provided the linear capacity of the adsorbent is not exceeded. Beyond this value R_F increases or decreases with sample size depending on whether the shape of the isotherm is convex or concave.

(xi) Complexity of Sample. The R_F value of a solute can be affected, i.e., displaced to higher or lower values, by the proximity of other solutes. For reference purposes the R_F values of pure compounds chromatographed singly should therefore always be quoted.

(xii) Accuracy of Measurement. With asymmetric spots difficulty will be encountered in determining where the true mass center of the spot should be. Fortunately, except with very polar and chemisorbed solutes, symmetric spots are usually obtained. The distances the spot and mobile phase move are usually measured with a ruler calibrated in millimeters. Each measurement will involve an error of ± 0.5 mm. For a chromatogram in which the spot has moved 5 cm and the solvent front 10 cm the error due to measurement will be 1%, i.e., the R_F value will be 0.50 ± 0.005.

6.8.3. Conclusions

As a result of all these variables it can be seen why, in the past, the reproducibility of R_F values in TLC has been so poor. Now that we know some of these variables progress can be made in further standardizing the technique and we have tried to indicate what the standard procedure should be.

In spite of this poor reproducibility, many of the tables of R_F values published in the literature are useful because: (a) they indicate whether it is possible to separate a particular group of compounds; this can be gleaned from a table of R_F values published by one particular laboratory using its own particular set of standard conditions: and (b) they serve as a useful guide to the conditions to choose when one is attempting to carry out a similar separation for the first time.

6.9. DETECTION OF SOLUTES

Colorless solutes on a developed chromatogram are detected either by viewing the plate under ultraviolet light or by revealing the spots as colored zones by spraying the plate with a chromogenic reagent. A third, more specialized method exists in which radioactive solutes are located by Geiger or gas flow counters.

6.9.1. Inspection under Ultraviolet Light

Plates are best viewed in a darkened room. Many suitable ultraviolet sources are commercially available in which two lamps are incorporated, one with an emission maximum at 254 nm and the other with a maximum at 365 nm, i.e., short- and long-wave lamps, respectively. Chromatograms revealed in this way can also be photographed.

On a pure adsorbent (e.g., Silica Gel H) fluorescent substances absorb long-wave ultraviolet light and re-emit this energy at a longer wavelength in the visible region. For example, polynuclear aromatic hydrocarbons appear as brightly colored spots on a dark background. The color (yellow, orange, green, blue, mauve) often aids in the identification of the material. This method can detect less than 0.1 μg of many fluorescent solutes.

An inorganic phosphor (e.g., zinc silicate or sulfide) incorporated in the adsorbent enables many ultraviolet-absorbing solutes to be seen as dull mauve spots on a bright green fluorescent background when viewed under short-wave ultraviolet light. The limit of detection by this method is about 5 μg.

6.9.2. Revealing Reagents

The usual way of locating colorless solutes on a developed chromatogram is to spray the plates with a solution of a reagent which will chemically react with the solute. Some reactions take place in the cold, others require the application of heat—even to the extent of promoting charring of the solute after spraying with an aggressive chemical. In this latter case the presence of organic binders or support materials would interfere and therefore cannot be used. Finally, some plates need to be viewed under ultraviolet light after spraying to evaluate the effects of the chromogenic reagent.

Spraying the plate is best carried out with a commercial sprayer of the type similar to that shown in Fig. 6.10. This consists of a 120-ml-capacity glass jar containing the reagent solution and a propellant gas container, the two parts being connected by a plastic junction piece containing tubing, valves, and a small orifice which delivers a fine spray on depressing the valve. The propellant is usually a mixture of fluorocarbons. All parts are cheap and interchangeable and even many corrosive reagents can be used with the all-plastic models. The fine spray ensures that the reagent can be uniformly applied to the plate. The spraying operation must be carried out in a fume cupboard with an efficient draft.

Fig. 6.10. Spray gun.

Chromogenic reagents may be broadly classified into two types: (a) general reagents, which react with a wide variety of different compound types, and (b) specific reagents, which indicate the presence of a particular compound or functional group. Examples of reagents of both types commonly used in industrial organic analysis will now be described.

(a) General Reagents

(i) Iodine Vapor. The plate is placed in a closed jar containing a few crystals of iodine. The iodine vapor dissolves in most organic compounds, which show up as brown spots on a pale yellow background. In many cases when the spots have been located the iodine can be evaporated and the plate resprayed with another reagent, or the solute can be eluted from the adsorbent for further analysis. In this case it should be noted that the chemical nature of the solute might have been altered, since many compounds undergo substitution or addition reactions with iodine. Spots revealed with iodine fade rapidly when left in the open atmosphere. This can be delayed and the reaction made more sensitive by subsequently spraying the plate with 7,8-benzo-flavone (0.3 g dissolved in 95 ml ethanol + 5 ml 30% sulfuric acid). This reagent produces an intense blue coloration with iodine. For maximum color contrast the plate should be sprayed after the iodine color of the background has just faded.

(ii) Dichromate–Sulfuric Acid. The reagent is prepared by dissolving 5 g of potassium dichromate in 100 ml of 40% sulfuric acid. Spraying the solvent-free plate followed by short heating at 110°C produces light blue spots on an orange background with most organic substances. Heating at higher temperatures (200°C) chars the spots, i.e., forms black zones on a colorless background. This technique cannot be used with adsorbents containing organic binders or with plastic or metal-supported precoated layers. Other oxidizing or aggressive chemicals, such as nitric acid, acetic anhydride, percholoric acid, or phosphoric acid, can be used with sulfuric acid instead of potassium dichromate.

(iii) Phosphomolybdic Acid. Spraying of the chromatogram with a 5% solution of the acid in isopropanol followed by a short heating at 110°C gives blue-black spots on a yellow background with a large variety of organic substances. Subsequent exposure of the plate to ammonia vapors decolorizes the background. The color is very stable and chromatograms can be stored for years in the dark. Well-separated spots are very suitable for quantitative analysis when revealed with this reagent. The reagent is available, ready for immediate use, from a number of supply houses in the form of an aerosol spray can.

(iv) Rhodamine B. Fifty milligrams of the dye are dissolved in 100 ml of ethanol. The reagent produces mauve spots on a pink background with a large number of organic substances. The spots are intensified by blowing sufficient bromine vapor onto the sprayed chromatogram to decolorize the background. Inspection of the plate under long-wave ultraviolet light shows orange fluorescent spots on a dark background.

(v) Antimony Trichloride or Pentachloride. The reagents are used as 10–20% solutions in chloroform or carbon tetrachloride. Heating the sprayed plates to 110°C for a short period produces spots of various colors on a white background with a wide variety of organic substances. Their differences in color may be used to characterize certain classes of compounds.

(vi) Water. Spraying with water until the layer appears translucent reveals hydrophobic solutes as white opaque spots on a semitransparent background. After outlining the spots with a sharp pencil the plate can be dried and either resprayed with another reagent, or the layer scraped with a microspatula and the solute eluted from the adsorbent for further analysis.

(b) Specific Reagents

Stahl[57] and Kirchner[58] list 266 and 204 spray reagents, respectively, to which the reader is referred. Many of these reagents have been designed to reveal natural products or biochemicals. In Table 6.2 we give a selection of reagents more likely to be used in industrial laboratories which we have found to be satisfactory.

Table 6.2. Specific Spray Reagents for Functional Groups

Compound class	Reagent	Procedure	Results
Acids	Bromocresol green	Dissolve 0.04 g in 100 ml ethanol; add 0.1 N Na-OH until blue color just appears	Yellow spots on green background
	2,6-Dichlorophenol-indophenol (sodium salt)	Dissolve 0.1 g in 100 ml ethanol	Red spots on blue background
Alcohols	Ceric ammonium nitrate	Dissolve 6 g in 100 ml $4N$ HNO_3	Brown spots on yellow background
	Vanadium oxinate	Dissolve 0.1 g in 100 ml alcohol-free chloroform	Red spots on green background
Aldehydes	o-Dianisidine	Saturated solution in acetic acid	Various
	2,4 Dinitrophenyl hydrazine	Dissolve 0.4 g in 100 ml $2N$ HCl	Yellow-to-red spots on pale orange background
		Respray with 0.2% potassium ferricyanide in $2N$ HCl	Olive green spots on yellow background.
Amides	Hydroxylamine-ferric nitrate	(a) Dissolve 1 g hydroxylamine hydrochloride in 9 ml water; (b) dissolve 2 g sodium hydroxide in 8 ml water; (c) dissolve 4 g ferric nitrate in 60 ml water and 40 ml acetic acid; spray with mixture of one volume (a) and one volume (b); dry at 110° C for 10 min; spray with mixture of 45 ml (c) and 6 ml conc. HCl	Variety colored spots on colorless background
Amines	Cobalt thiocyanate	Dissolve 3 g ammonium thiocyanate and 1 g cobalt (II) chloride in 20 ml water	Primary, secondary and tertiary aliphatic amines give blue spots on pinkish-white background
	Indanetrione hydrate (ninhydrin)	Dissolve 0.3 g in 100 ml n-butanol and add 3 ml acetic acid	Yellow, pink, or mauve spots on white background
	Salicylic acid	Dissolve 1 g in 100 ml toluene	Primary, aliphatic, and aromatic amines give yellow spots on white background

Table 6.2 *(Continued)*

Compound class	Reagent	Procedure	Results
	p-Dimethylamino-benzaldehyde	Dissolve 1 g in 100 ml acetic acid	Aromatic amines give yellow or orange spots on colorless background
	2,3-Dichloro-1,4-naphtho-quinone	Dissolve 1 g in 100 ml toluene	Most amines give orange-to-purple spots on light orange background
	Diazotized *p*-nitroaniline	Dissolve 0.3 g *p*-nitrobenzene diazonium fluoroborate in 35 ml dioxane and 65 ml water	Aromatic amines give yellow, orange, or red spots on pale yellow background
Anhydrides	Hydroxylamine-ferric nitrate	See Amides	—
Antioxidants	Dichloroquinone-4-chlorimine (or corresponding dibromo compound)	Dissolve 1 g in 100 ml ethanol; respray after 15 min with 2% borax in 4% ethanol	Various colors on white background
Bases	Bromocresol green	See Acids	Blue spots on green background
Chlorinated hydrocarbons	Silver nitrate	Dissolve 0.1 g silver nitrate in 1 ml water, 10 ml 2-phenoxyethanol, and 190 ml acetone; add one drop 30% hydrogen peroxide; after spraying expose 15 min to long-wave UV light	Grey spots on colorless background
Chlorine-containing insecticides	Methyl yellow	Dissolve 0.1 g in 75 ml ethanol and 25 ml water; after spraying expose 5 min to UV light	Red spots on yellow background
	Diphenylamine–zinc chloride	Dissolve 0.5 g of each in 100 ml acetone; heat 5 min at 200°C	Various colors
	N,N-dimethyl-*p*-phenylene-diammonium dichloride	Dissolve 0.5 g in 100 ml ethanol in which 1 g sodium has previously been dissolved; after spraying expose 1 min to UV light	Red spots on pale background

Table 6.2 *(Continued)*

Compound class	Reagent	Procedure	Results
3,5-Dinitro-benzoate esters	α-Naphthylamine	Dissolve 1 g in 100 ml ethanol	Orange spots on white backgound
1,2-Diols	Lead tetraacetate	Dissolve 1 g in 100 ml toluene	White spots on brown background
Ethanol-amines	Benzoquinone	Dissolve 1 g in 20 ml pyridine and 80 ml n-butanol	Red spots on pale background
Esters	Hydroxylamine–ferric nitrate	See Amides	
Hetero-cyclics	Diazotized-p-nitroaniline	See Amines	
	Formaldehyde–sulfuric acid	See Polynuclear aromatics	
	Tetracyano-ethylene	Dissolve 10 g in 100 ml toluene	Various colors on white background
Hydroxamic acids	Ferric–chloride	Dissolve 2 g in 100 ml 0.5 N HCl	Red spots on colored back-ground
Ketones	o-Dianisidine	See Aldehydes	Various colors
	2,4-dinitrophenyl hydrazine	See Aldehydes	Blue spots on white background
Methylene groups (activated)	Sodium nitroprusside	Dissolve 1 g in 50 ml 2 N NaOH and 50 ml ethanol	Red-to-violet spots on pale back-ground
Nitrate esters	Diphenylamine	Dissolve 1 g in 100 ml ethanol; after spraying expose to short-wave UV light	Yellow-green spots on white back-ground
Peroxides	N,N-Dimethyl-p-phenylene diammonium dichloride	Dissolve 1 g in 75 ml methanol, 24 ml water, and 1 ml acetic acid	Purple spots on pink background
	Ferrous thiocyanate	(a) Dissolve 4 g ferrous sulfate in 100 ml water; (b) dissolve 1.3 g ammonium thiocyanate in 100 ml acetone; spray with a mixture of 10 ml of (a) and 15 ml of (b)	Red-brown spots on pale background
Phenols	4-Amino antipyrine	(a) Dissolve 3 g in 100 ml ethanol; (b) dissolve 8 g potassium ferricyanide in 100 ml H₂O; spray with (a), then (b), and then expose to ammonia vapor	Red, orange, or pink spots on pale background

Table 6.2 *(Continued)*

Compound class	Reagent	Procedure	Results
	2,6-Dichloroquinone-4-chlorimine	See Antioxidants	Various colors on white background
	Ferric chloride	See Hydroxamic acids	Blue or green spots on white background
	Diazotized *p*-nitroanaline	See Amines	Yellow-orange-brown spots on colorless background
	Tetracyanoethylene	See Heterocyclics	—
	Vanillin	Dissolve 1g in 100 ml conc. H_2SO_4	Yellow spots on white background
Phenothiazines	Palladium chloride	Dissolve 0.5 g in 50 ml water and 50 ml acetone add two drops conc. HCl	Blue spots on pale yellow background
Phosphate esters	Ammonium molybdbate–perchloric and	Dissolve 0.5 g ammonium molybdate in 5 ml water and 1 ml conc. HCl; add 2.5 ml perchloric acib and cool; dilute to 50 ml with acetone and let stand 24 hr; after spraying expose for 2 min to an IR lamp, then 7 min to long-wave UV light	Blue psots on white background
Phosphoric acids	Ammonium molybate–stannous chloride	(a) Dissolve 1 g ammonium molybdate in 100 ml water; (b) dissolve 1 g stannous chloride in 100 ml 10% HCl; spray with (a), dry, then spray with (b)	Blue spots on white background
Phosphorus-containing insecticides	4-(*p*-Nitrobenzyl) pyridine	(a) Dissolve 2 g in 100 ml acetone; (b) dissolve 10 g ammonium carbonate in 50 ml water and 50 ml acetone; spray heavily with (a) and then lightly with (b)	Blue spots on white background
Phthalate esters	Diazotized *p*-nitroaniline	See Amines	Yellow spots on white background
	Resorcinol	(a) Dissolve 2 g zinc chloride and 20 g resorcinol in 100 ml ethanol; (b) 4 N H_2SO_4; (c) 40% KOH; spray with (a) and heat 10 min at 150° C; spray with (b) and heat 20 min at 120°C; finally, spray with (c)	Orange-red spots on yellow background

Table 6.2 *(Continued)*

Compound class	Reagent	Procedure	Results
Polynuclear aromatic hydro-carbons	Formaldehyde–sulfuric acid	Dissolve 2 ml 37% formaldehyde solution in 100 ml conc. H_2SO_4	Various colors on white background
	Tetra-cyanoethylene	See Heterocyclics	Various colors on white background
Sulfate, sulfonate esters	Pinacryptol yellow	Dissolve 0.05 g in 100 ml water; view under long-wave UV light	Orange spots on dull blue background
Sulfur-containing compounds	Palladium chloride	See Phenothiazines	Brown spots on pale yellow background
	Iodine azide	Dissolve 3 g sodium azide in 100 ml 0.1 N iodine; discard immediately after use (*SOLID IODINE AZIDE IS EXPLOSIVE!*)	Deep brown spots on pale brown background
Terpenes	Diphenyl picryl hydrazyl	Dissolve 60 mg in 100 ml chloroform	Yellow spots on purple background
Thiophosphate esters	Palladium chloride	See Phenothiazines	Brown spots on pale brown background
Unsaturates	Fluorescein–bromine	Dissolve 0.1 g fluorescein in 100 ml ethanol; after spraying expose to bromine vapor; view under longwave UV light	Yellow spots on pink background

6.10. CHROMATOGRAPHIC RECORDS

In recording the details of a thin-layer chromatogram the following information is required: type of sample, its size, and concentration; description of stationary phase, layer thickness, mobile phase, development chamber, and its saturation state; and finally, the revealing method used, its sensitivity, and the R_F values of the separated components. A list of R_F values, however, does not give a true picture of the chromatogram because it does not describe the shape and size of the spots, i.e., no information is given on the achievement of separation, nor does it give any guide on the quantitative composition of the sample, i.e., whether a solute is present as a major, minor, or trace component. To keep a record of this the actual chromatogram, or a copy, must be preserved. Alternatively, an analogous record may be obtained, e.g., a densitometric trace.

Merck precoated plates or the less bulky Chromogram sheets can conveniently be preserved in laboratory notebooks or in a darkened drawer because they possess strongly adhering adsorbent layers. Chromatographic details can be lightly written on the adsorbent film with a soft pencil or even a ball-point pen. If the spots are visible only under ultraviolet light their boundaries can also be pencilled-in directly on the layer. The revealing reagent used must produce stable colors. Phosphomolybdic acid is a good general revealing reagent for this purpose and we have preserved for several years chromatograms sprayed with this reagent.

In cases where the colors are expected to fade, the chromatogram must be copied. In the simplest case the chromatogram is traced onto transparent paper. Alternatively, it can be conveniently photographed with a Polaroid camera. If the chromatogram is particularly valuable, then it is worthwhile photographing either to produce a color transparency to be made up into a slide, or to produce a color negative from which any amount of color positive prints can be made. Photographs can either be taken under daylight or ultraviolet light with appropriate filters. For the best results the incident light should strike the plate at 45°. Chromatograms with poor contrast are best photographed by transmitted light.

6.11. DIRECT QUANTITATIVE ANALYSIS ON THE LAYER

6.11.1. Introduction

Quantitative evaluation of thin-layer chromatograms can be divided into two categories. In the first the solutes are assayed directly on the layer, while in the second they are eluted from the adsorbent before being examined further.

Assay on the layer can be carried out by measuring the area alone or, better, by measuring the integrated spot area and density on a developed and revealed chromatogram or on photocopies or photographs of revealed chromatograms. These values are then related to the amount of substance in the spots through the use of standards and calibration curves. These measurements can be made by visual comparison of samples and standards chromatographed simultaneously, by direct measurement of areas, by densitometry, fluorimetry, or spectrophotometry, and by counting radioactive isotopes.

In direct quantitative analysis on the plate certain conditions must be fulfilled in the production of the chromatogram. This is because the area of a spot formed by a particular solute in a developed chromatogram depends upon the polarity of the solvent used to make up the solution to be chromatographed, the distance the spot travels on the chromatogram, and the proximity of other spots on the chromatogram. The following conditions must therefore be fulfilled before attempting a direct quantitative analysis on the layer.

(a) The same volume of solutions of the samples and standards must be spotted on the plate. The area of the starting spot must be the same in each case. This is achieved by adding the solution drop by drop, and allowing the solvent to evaporate after the addition of each drop.

(b) The solvent must be the same for both samples and standards.

(c) Samples and standards must be chromatographed side by side on the same plate to minimize variations in layer thickness and the revealing technique.

(d) The concentrations must be adjusted so that the standards and samples are closely bracketed, and also so that the response is maximized and linear. This will depend on the nature of the solute and the sensitivity of the the revealing reagent. In general the smallest standard should be just visible and the highest standard should be about five times the minimum concentration. The most accurate results will be obtained on those samples that lie on the lowest part of the calibration curve.

(e) The distance traveled by all standards and samples on the same plate must be identical. This is best achieved by using sandwich chambers.

The techniques to be used in applying the samples and obtaining the chromatogram have already been described under the appropriate section headings. We have found Merck precoated silica gel plates satisfactory for direct quantitative analysis on the layer. They have uniform, tightly bound layers which can withstand the extra handling which is unavoidable in an analysis of this type. Because of the very fine particle size of the silica, the mobile phase velocity is less than with hand-coated plates and compact uniform spots are readily obtainable. Although sulfuric acid charring techniques cannot be used because of the organic binder present, most other revealing reagents can be used, including some aggressive reagents such as phosphomolybdic acid and antimony trichloride.

The fact that TLC can be a rapid and simple technique using only inexpensive apparatus is of overriding importance to some laboratories. Such laboratories will be content to use the visual comparison technique and will be satisfied with a precision that can lie anywhere within the range 5–30% of the amount present. Other laboratories, needing to get the most out of their chromatograms, will avail themselves of one or more sensitive measuring devices which will enable quantitative data within the precision range 2–10% to be achieved. Each of the various direct quantitative methods will now be considered in turn.

6.11.2. Visual Comparison

The advantage of this method is that it is quick, requires no equipment, and is satisfactory if only a semiquantitative estimate ($\pm 20\%$) is needed.

The precision can be improved to better than $\pm 10\%$ for repetitive, routine analysis of samples of similar composition if strict adherance to the conditions noted above are observed. Greatest accuracy is achievable where the samples and standards lie in the concentration range where the intensities and areas of the spots are changing most rapidly with changes in the weight of solute. Because of the effect neighboring solutes have on each other, the composition of the reference solutions must match the sample as closely as possible. For increased precision it is desirable to have the same plate independently evaluated by several workers. In this context it is worth mentioning that the young feminine eye can be more discerning than that of the aging male. An obvious, but often neglected requirement is that the observer should not be color blind.

6.11.3. Area Measurement

This measurement, although tedious and time-consuming, is more objective than the visual comparison method and produces results with an error of 5–10%. However, painstaking care must be taken in producing a chromatogram because of the many factors, already discussed, that influence the area of the final spot. Changes in spot intensity are not accounted for by this method. Spot areas can be measured by planimetry, copying onto tracing paper and weighing, photographing and weighing, or by copying onto millimeter-squared paper and counting the number of squares. Difficulties arise in deciding where the boundary of the spot occurs. Since in general TLC spots are small (15–150 mm^2), this is a major source of error.

The amount of material in the sample spot is found by interpolation after plotting some function of the weight of material in the standard spots against some function of the spot areas. For example, Giddings and Keller[25] have shown theoretically, and Brenner and Niederwieser[26] practically, that a linear relationship exists between spot areas and the log weight of the applied solute. On the other hand, Purdy and Truter[27] found a linear relationship between the square root of the spot area and the log weight of the applied solute, and that this relationship held true over a wider concentration range.

To avoid plotting calibration curves, these latter authors proposed an algebraic method based on three analyses on the same plate, namely a standard solution, the sample solution, and a diluted sample solution. The weight W of solute in the sample solution is then given by

$$\log W = \log W_s + [(\sqrt{A} - \sqrt{A_s})/(\sqrt{A_d} - \sqrt{A})] \log D$$

where W is the weight of solute in the sample solution (before dilution), W_s is the weight of solute in the standard solution, D is a dilution factor, A_s is the spot area of the standard, A is the spot area of the solute in the sample solution, and A_d is the spot area of the solute in the diluted sample solution.

Using this expression Purdy and Truter found an average standard deviation of 3.1% for 600 separations based on adsorption and 3.6% for 980 separations based on partition. Use of this method has been reported many times in the literature since then, standard deviations ranging between 5 and 10%.

6.11.4. Densitometry

In contrast to the methods discussed so far, a 20 × 20 cm TLC plate containing five standards and ten samples can be evaluated by densitometry in less than 10 min, but it requires the purchase of a more expensive instrument.

In this technique an integrated function of spot area and intensity is measured. This means that the initial area of the applied spot is not quite so critical as in the previous two methods, though considerable care should still be taken. The same conditions affecting the area of the final spot also affect its intensity, but in the reverse sense. Goldman and Goodall([28]) have discussed the theory of densitometry based on the Kubelka–Munk theory, which describes the interaction of light with diffuse absorbing surfaces. The results show that transmission measurements are preferable, on theoretical grounds, to reflection measurements.

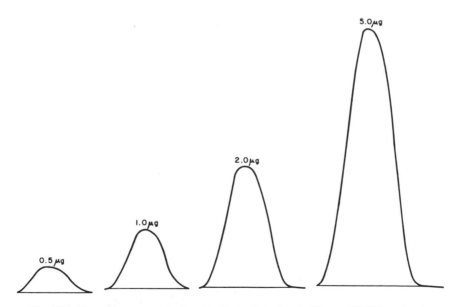

Fig. 6.11. Densitometer trace of antioxidant. Adsorbant, silica gel F 254; solvent, toluene; spray, phosphomolybdic acid; aperture, 10 × 0.5 mm; filter, blue.

The method involves scanning the final spot with a beam of light; either the reflected or transmitted light falls onto a photomultiplier. The difference in intensity between the incident light and the transmitted or reflected light is measured electrically and indicated on a chart as shown in Fig. 6.11. The height of the peak is a measure of the intensity of the spot, while the peak width is proportional to the spot length in the direction of scanning. By measuring the area under the curve, e.g., by measuring peak height times peak width at half-height or by using a built-in integrator, relationships between peak area and the weight of solute in the spots are obtained. An example is shown in Fig. 6.12. It has been found that over limited concentration ranges a linear relationship exists between peak area and the square root of the weight of solute. All spots must be on the same chromatogram and produced with all the precautions we have already described. It is probably true to say that the errors involved in producing a reproducible

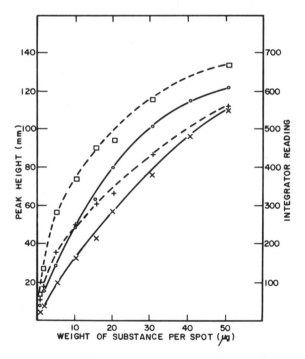

Fig. 6.12. Calibration curves for *p*-nonyl phenol; (□) peak height, diazotized *p*-nitroaniline; (+) integrator, diazotized *p*-nitroaniline; (○) peak height, phosphomolybdic acid; (×) integrator, phosphomolybdic acid. (Reproduced from[56] by courtesy of Springer-Verlag.)

Table 6.3. Repeatability of Measurement of a Single Spot[a]

	Peak height measurement	Integrator measurement
Number of determinations	18	8
Mean, mm	122.6	573
Range, mm	121.0–124.0	560–585
Standard deviation	1.0	7.9
Percentage error	±0.8	±1.4

[a] A total of 15 μg p-nonyl phenol, revealed with diazotized p-nitroaniline, in a lubricating oil.

Table 6.4. Precision Data for the Determination of p-Nonyl Phenol in Lubricating Oil

	Integrator measurement	Peak height measurement
Number of determinations	10	10
Mean value, μg	15.0	15.4
Actual value, μg	15.0	15.0
Range, μg	14.6–15.4	14.6–15.8
Standard deviation	0.26	0.41
Percentage error	±1.7	±2.8

Table 6.5. Commercial Densitometers

Name	Manufacturer
Joyce-Loebl Chromoscan with Thin Layer Attachment	Joyce-Loebl & Co., Ltd. Team Valley, Gateshead County Durham, England
Photovolt Densitometer Model 300	Photovolt Corp., 1115 Broadway, New York, N.Y. 10010
Camag-Turner Fluorometer	Camag A.G. Muttenz, Switzerland
Vitatron Densitometer TLD 100 Flying Spot Scanner	Vitatron N.V. Spoorstraat 23, Diesen, The Netherlands
Aminco Motorized Thin Film Scanner	American Instrument Co. Silver Spring, M. 20910.
Nester-Faust Uniscan 900	Nester/Faust Manufacturing Corp. 2401 Ogletown Road Newark, Del. 19711
Schoeffel Model SD-3000	Schoeffel Instrument GmbH. 2351 Trappenkamp, C. Strasse 2, W. Germany
Vis-UV Chromatogram Analyzer	Farrand Optical Co., Inc., 535 South 5th Ave, Mount Vernon, N. Y. 10550
Chromatogram Spectrophotometer PMQ	Carl Zeiss, Oberkochen/Wuert., W. Germany

chromatographic spot far outweigh the errors inherent in the densitometers themselves (compare Table 6.3 and 6.4).

Although several instruments are now available (Table 6.5), at the present time the two most commonly used seem to be the Photovolt densitometer in the USA and the Chromoscan in Europe. A description of the Chromoscan is given in Ref. 29. and critical evaluations of its applicability to TLC given in Refs. 30 and 31. It has been employed by a number of workers with considerable success([32-36]) using various revealing reagents, e.g., sulfuric acid charring, phosphomolybdic acid, ninhydrin for amino compounds, diazotized p-nitroaniline for phenols and iodine (the plate is covered by a thin sheet of glass to slow down the evaporation rate of the iodine).

The largest single source of error in densitometry lies in revealing the spots. First, a reagent must be chosen which produces a colored spot contrasting sharply with the background; second, the intensity of the color must be reproducible and be proportional to the amount of substance present; third, the plate must be evenly sprayed to produce a uniform, pale background. The contrast of the spot may be enhanced by inserting a filter of complimentary color in the incident light beam.

Spots are usually scanned in the direction of development of the chromatogram provided they are sufficiently well resolved to allow the position of the baseline (corresponding to the background color of the sprayed plate) to be ascertained. Greater sensitivity is obtained by scanning with a spot source, but because it is more difficult to align the spot source with the center of the TLC spot, greater accuracy is obtained by scanning with a slit source whose width is less than the diameter of the TLC spot.

The reflectance mode of operation has to be used with thick layers or with nontransparent supports and this measurement also includes a certain amount of scattered light. In the transmittance mode, however, scattered light is not measured and very satisfactory results are obtained using Merck's precoated silica gel glass plates, as shown in Tables 6.3 and 6.4. Plates must first be checked in the densitometer to ensure they are of uniform thickness. In general, densitometry gives quantitative results with standard deviations in the 2–5% range.

6.11.5. Fluorimetry

This term covers (a) fluorescence measurement in which a fluorescent spot on a dark, nonfluorescent background is scanned by an ultraviolet light source having an emission maximum, usually at 366 nm, and (b) fluorescence quenching, in which an ultraviolet-absorbing compound, appearing as a dark spot on a bright, fluorescent background, is scanned by an ultraviolet light source having an emission maximum at 254 nm. The preparation of the chromatogram, the mechanics of the scanning, and the interpretation of the results are similar to those described for densitometry.

The determination of substances having a natural fluorescence, such as polynuclear aromatic hydrocarbons, can be reasonably accurate with a standard deviation of 2%. The main source of error here is due to inaccuracies in sample application, since use of the unprecise spraying technique is avoided. The limit of detection is very low (of the order of 100 ng for certain solutes). Some compound classes have to be sprayed with a fluorogenic reagent, such as Pinacryptol Yellow for sulfonates, before they can be measured as fluorescent spots. The inclusion of a spraying step pushes the standard deviation up to 5%[37].

The precision of fluorescence quenching is not as good as with the straightforward measurement of fluorescence, standard deviations of 4–7% having been reported[38,39]. The sensitivity of quenching measurements is poorer than that achieved by many conventional revealing reagents, and is of the order of 5–10 μg. As with fluorescence, two cases can be distinguished according to whether quenching occurs with or without spraying with a fluorogenic reagent[40], e.g., Rhodamine B.

Among the instruments commercially available for direct fluorimetry of TLC plates is the Camag-Turner TLC Scanner (Camag, Muttenz, Switzerland), the description and performance of which has been given by Jänchen and Pataki[37].

6.11.6. Reflectance Spectrophotometry

A natural extension of densitometry and fluorimetry is the capability or recording ultraviolet-absorption, ultraviolet-fluorescence, and visible spectra of separated solutes directly on the layer. An advantage of this technique is that two or more unresolved compounds, e.g., stereoisomers, can be determined in the presence of each other if their spectra are sufficiently different. Usable spectra can be obtained on as little as 1–5 μg of solute and therefore this technique is at least ten times more sensitive than the corresponding determination after the solute has been eluted from the adsorbent.

Frodyma et al.[41] describe the construction and use of a suitable cell into which the spot adsorbed on the sorbent can be placed, for use with a Model DK-2 spectrophotometer fitted with a standard reflectance attachment.

An automatic recording instrument (Table 6.5) has been designed by C. Zeiss in conjunction with Stahl and Jork[42] which records spectra in the 200–2500-nm range, i.e., from the ultraviolet to the near-infrared.

6.11.7. Radioactive Methods

Radioactive isotopes are often used to follow the course of a chemical reaction, and their qualitative and quantitative distribution in a reaction mixture can be examined by TLC. A general account of isotope techniques in TLC has been given by Mangold[59]. There are three main ways of detecting radioactive isotopes on TLC plates.

(i) Autoradiography. A piece of film sensitive to α-, β-, γ-, or X-rays is laid on top of the developed chromatogram. Radioactive spots appear as dark zones on the developed film, which is then evaluated densitometrically. The major disadvantage of the method is that exposure times can vary from a few hours to several weeks, depending on the level of radioactivity.

(ii) Direct Scanning. The chromatogram is scanned by a radiation-sensitive device such as a gas-flow proportional counter or a Geiger–Müller counter to produce a record similar to a densitometer trace. An example of an instrument based on the gas-flow detector and including automatic transport of the place and recording of quantitative data is the Thin-Layer Scanner constructed by Schulze and Wenzel[43] and marketed by Desaga. This apparatus has been evaluated by Wilde[44] and the effect of instrumental variables has been studied by Wood[60] who showed that, for optimum results, the main requirements were (i) a low scan speed, to enable the maximum number of counts to be recorded; (ii) a narrow slit, for maximum peak separation; and (iii) the minimum distance between chromatogram and detector, to improve resolution and detector efficiency. No typical precision data on quantitative direct radioactive scanning are presently available. According to Mangold, it is an insensitive technique because of the high self-absorption of the radiation and the "softness" of the (glass) background; it is therefore unreliable as a quantitative procedure. One advantage is that after scanning for radioactivity the chromatogram is available for spraying with a conventional chemical reagent.

(iii) Liquid Scintillation Counting. In this technique a portion of the adsorbent is scraped off into a vial (this can be done automatically with a zonal scraper) [45] and mixed with a scintillator solution and the suspension is then counted in a scintillation counter. The resolution depends upon the width of adsorbent taken for the measurements, and the disadvantage of the method is the large number of samples that have to be taken if the whole length of a chromatogram is to be examined.

6.12. REMOVAL OF SOLUTE FROM ADSORBENT LAYER

6.12.1. Introduction

The combination of liquid chromatography with other sensitive instrumental techniques provides a powerful weapon for the identification of complex mixtures. Examination of fractions from chromatographic separations may be needed to satisfy any of the following three requirements: further separation, quantitative analysis, and identification of separated solutes.

The effectiveness of a chromatographic separation improves as the sample size is reduced, but, fortunately, sensitive analytical instrumentation has now been developed to a stage where very small quantities of material can be effectively handled. Some typical sample requirements of various analytical methods are given in Table 6.6, which shows the compatibility of the chromatographic techniques with each other and with several instrumental techniques.

The problems associated with combining techniques of this nature can be classified as follows: avoidance of solute contamination, incomplete recovery, and physical handling of small samples. These problems, although common to both column and layer techniques, are more manifest in TLC and the following discussion will relate chiefly to TLC. Their relevance to column chromatography will be evident.

The major problem is to eliminate contamination. The following example gives an idea of the magnitude of the problem. To obtain an infrared spectrum on 10 μg of solute (commercial instruments and technical skills for doing this are now fully developed) means that not more than 0.5 μg of impurity can be tolerated (i.e., 5% contamination). This means that the solute must be completely separated from a pure adsorbent by eluting with not more than 0.5 ml of a volatile solvent containing less than 1 mg liter^{-1} of nonvolatile matter. Furthermore, the solute has to be handled in such a

Table 6.6. **Typical Sample Requirements of Various Anayltical Methods (in μg)**

Thin–layer chromatography (adsorption)	20
Liquid chromatography in columns (adsorption)	1000
Gas–liquid partition chromatography (packed columns)	100
Gas–liquid partition chromatography (capillary columns)	1
UV spectroscopy (absorption)	50
UV spectroscopy (fluorescence)	1
Mass spectrometry (direct insertion probe)	20
Nuclear magnetic resonance spectroscopy	10,000
Infrared spectroscopy (micro KBr disk)	20
Molecular weight determination	10,000
C,H,N analysis	1,000
Atomic absorption spectroscopy	10
X-ray fluorescence spectroscopy	10,000
Polarography (pulse)	1
Microcoulometry (S,N)	1

way so as not to pick up any further impurities from the hardware or laboratory atmosphere until its spectrum has been recorded.

6.12.2. Adsorbent Purity

Recent work([46]) has shown that the presence of organic impurities in silica gel is just as prevalent today as it was when first investigated in 1962 by Honegger([47]). During this time the consequences of these impurities have been manifest in scores of papers (e.g., Ref. 48), especially those utilizing sensitive spectroscopic techniques or gas chromatography. Organic impurities in adsorbents can be introduced (a) during the manufacturing process, (b) from the plastic containers in which many adsorbents are supplied, and (c) from the atmosphere. Inorganic impurities are almost always present. For example, iron is present in most silicas. The situation is further complicated by the inclusion of additives, both organic and inorganic, to improve the cohesive properties of the layer, to modify its sorption properties, or to act as indicators. We have found Silica Gel HR to be a sufficiently pure adsorbent to use with all the ancillary techniques we have tried, provided it is stored in a clean glass container and protected from the laboratory atmosphere. Silica Gel HF254 can also be used in cases where possible contamination by the inorganic phosphor presents no problems. The presence of the phosphor, of course, aids in locating the position of the spots on the chromatogram. Silica Gel HF254 is, however, supplied in plastic containers. Each batch therefore has to be checked for the presence of organic impurities, and, if satisfactory, transferred to and stored in a tightly stoppered glass bottle.

All the precoated plates and sheets tested so far have contained substantial amounts of organic impurities. Plastic sheets are particularly unsatisfactory since plasticizers in the support tend to adsorb onto the adsorbent.

6.12.3. Solvent Requirements

To elute the solute from the silica gel, a solvent of the appropriate purity, properties, and eluent strength (see Table 3.5) must be chosen. The solute must of course be soluble in the solvent and not react with it. The correct eluent strength will be that in which the solute would have an R_F value of 0.8 or greater on the particular adsorbent–mobile phase combination. The properties and purity of the solvent depend to some extent on the particular ancillary technique to be used. In ultraviolet work, for example, the solvent must be ultraviolet-transparent, while the identification of solutes by infrared spectroscopy and mass spectrometry requires nonvolatile matter to be absent. Even many analytical-grade solvents contain large (i.e., greater than 1 mg liter^{-1}) amounts of nonvolatile matter and consequently have to be distilled in glass immediately before use. Two commonly re-commended solvents which nevertheless have given trouble are chloroform (many batches of which, from several sources, were found to contain plas-ticizers) and ethers (diethyl, di-isopropyl, tetrahydrofuran, which contain inhibitors to prevent peroxide formation). Alcohols extract fine silica gel particles to form a colloidal dispersion, and therefore should not be used at this stage for spectroscopic work.

A problem frequently encountered is incomplete elution of the solute from the adsorbent. In repetitive, routine quantitative analysis by ultraviolet, for example, this does not matter so much provided the recoveries are reproducible and that reference solutions are treated, and behave, in a similar manner. In qualitative analysis, however, there is a need to account for the poor recovery. Possible causes are (a) inaccurate sample application, (b) chemisorption, (c) the presence of other, unsuspected components in the mixture that have not been accounted for, (d) losses due to volatility of sample components, and (e) poor sample handling technique.

6.12.4. Sample Handling Technique

All operations must be carried out in a clean atmosphere. The first prerequisite is absolute cleanliness of all the hardware involved. The most effective way of cleaning glassware has been found to be a 30-min immersion in a 5% aqueous solution of an organic detergent in an ultrasonic bath, followed by thorough washing with distilled water and drying in a clean, dust-free oven. Chromic acid is not an effective cleaning agent in this application.

The nature of the sample and the amount of solute required will deter-mine whether a normal analytical-scale separation or a preparative-scale operation is required. Factors to be borne in mind are that resolution deteriorates as the sample size is increased, and that the solute under examination must be well separated.

Location of the solute can present problems. However, by viewing a plate by reflected light, it is surprising how many "colorless" solutes can be seen as darkened or opaque zones. Fluorescent solutes can be detected by viewing under long-wave (3660A) ultraviolet light, while ultraviolet-ab-sorbing solutes can be located by their quenching effect on fluorescent layers of Silica Gel HF254 when viewed under short-wave (2540A) ultraviolet light. If all these methods fail, a reference sample should be chromatographed at one end of the plate. The remainder of the plate is covered while the reference sample is sprayed with a chromogenic reagent–preferably one that reacts immediately at room temperature. A 1% chloroform solution of iodine is a useful general reagent for this purpose. Care must be taken, in developing the chromatogram, that the "edge" effect is avoided, otherwise the R_F values of the spots in the middle of the plate are lower than those in the reference sample. Development should therefore be carried out in sandwich chambers.

If the sample has been applied as a band, then the adsorbent-impregnated solute is best removed by means of the "vacuum-cleaner" device, many variations of which are described in the literature. A suitable variant is shown in Fig. 6.13.

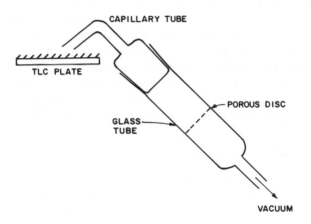

Fig. 6.13. Device for removal of adsorbent from TLC plate.

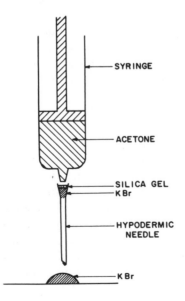

Fig. 6.14. Apparatus for transferring sample spot from adsorbent to potassium bromide. (Reproduced from[46] by courtesy of Elsevier Publishing Company.)

The procedure used for the isolation and identification of small(< 10 μg) samples is as follows: 40 μl of a 5% solution of the sample is applied as a short band, 2 cm in length, to a 5 × 20 cm TLC plate coated with Silica Gel HR, 750 μm thickness; 2 μl of a 1% solution of the same sample is chromatographed alongside and used as a reference. The chromatogram is developed in a sandwich chamber using a mobile phase containing less than 1 ppm nonvolatile residue and the 2-μl sample is sprayed with iodine solution so that the positions of the separated bands in the larger sample can be located. The band of solute under examination is scraped off with a micro-spatula and placed on top of a small amount of KBr powder and tamped down the cone joint of an 18 gauge hypodermic needle (see Fig. 6.14) The KBr serves as a filter to prevent silica gel fines passing through the needle. A 1-ml glass syringe is filled with pure acetone and connected to the needle and the solute is eluted drop by drop. About 20 drops are usually sufficient to elute the whole of the solute.

6.13. USE OF ANCILLARY TECHNIQUES

6.13.1. Spectrophotometry

This is the most commonly used quantitative method of TLC because a spectrophotometer is a standard piece of equipment found in most laboratories. Compared with densitometry, it is a much longer and more tedious procedure, requiring great care to ensure that impurities are eliminated and repeatable recoveries obtained. Nevertheless, many papers report standard deviations in the 3–5% range, and some are even better than this. A number of workers have compared spectrophotometry with densitometry and conclude that the precision of the two methods is similar[49, 50].

The initial sample size is chosen after considering the sensitivity (i.e., extinction coefficient, cell path length) and concentration range within which the Beer–Lambert law is obeyed. The chromatogram is developed in the standard way, the solute located, and the adsorbent transferred to one of elution devices described in Section 6.12.4, depending on the amount of adsorbent. The solute is eluted with a pure, volatile solvent into a 5-ml, pear-shaped flask, where the solvent is removed by gentle evaporation on a steam bath under a stream of nitrogen. The solute is redissolved in a solvent that does not interfere with the subsequent spectrophotometric determination. It is then transferred, using a hypodermic syringe, to a volumetric flask, a chromogenic reagent added if necessary, and made up to volume. For details of colorimetric procedures refer to standard treatises on spectrophotometric analyses, e.g., Ref. 51. The concentration of the separated solute in the original sample is obtained after establishing the relationship between absorbance and weight of solute in reference solutions.

An advantage of the spectrophotometric technique is that the shape of the absorption curve provides supporting evidence for the identification of the solute. The main disadvantages of the method arise from errors due to impurities (mainly from the adsorbent), incomplete recoveries, and the fact that the method is long and tedious. Since impurities absorb mainly in the ultraviolet region, measurements in the visible region are generally more reliable. The technique of forming a visible adduct on the plate by spraying, which is more easily dissolved than the original solute is not recommended because the spray reagent may not penetrate through the adsorbent to react with the whole of the solute.

6.13.2. Ultraviolet Fluorescence

The transfer technique used is as for ultraviolet absorption, but because of the greater sensitivity of the technique, special attention must be given to the elimination of fluorescent impurities. Because the sample requirements are so small, normal analytical-scale separations are used, with the advantage that samples and reference solutions can be run on the same plate. A blank solution from the adsorbent can be obtained by scraping off a small amount of silica gel near one of the sample spots and putting it through the whole elution procedure. Sample handling is minimized by using the microsyringe technique. Standard deviations reported in the literature range from 2 to 15%, probably reflecting difficulties in eliminating impurities.

The technique has been much used by Sawicki and his coworkers[52] in studying atmospheric pollutants using an Aminco-Bowman spectrophotofluorimeter.

6.13.3. Other Chromatographic Techniques

The separation of a complex mixture may require repetitive fractionation using different separation mechanisms. A useful first step is a rough fractionation into different functional groups using thin-layer adsorption chromatography, making full use of the high sample capacity of adsorbents. Selected zones of solute are scraped off, isolated from the adsorbent, and further separated, by gas–liquid partition chromatography if the components are volatile, or by high-efficiency liquid chromatography. Polymers can similarly be studied by exclusion chromatography or by pyrolysis–gas chromatography.

The reverse process, e.g., the use of TLC after gas chromatography, has been used by several groups of workers, but probably is not so generally useful. For a general account see Kaiser[61]. A TLC plate is placed 1 mm directly beneath the exit port of a gas chromatograph to which a hypodermic needle has been attached. The TLC plate can be programmed to move so that the individual peaks of the gas chromatograph are deposited on a different part of the TLC plate, which is then handled in the conventional manner. Separated solutes are thus characterized by retention indices as well as by R_F values and color reactions. However, the efficiency of the transfer operation from the gas chromatograph to the TLC plate is generally poor (30–80%). A recent application[53] involves the TLC separation of condensation products (2,4-dinitrophenyl hydrazones, semicarbazones), formed directly on the layer, of volatile carbonyl compounds issuing from the exit part of a gas chromatograph. Similarly, alcohols were studied by forming 3,5-dinitrobenzoates or o-nitrophenyl urethanes.

6.13.4. Infrared Spectroscopy and Mass Spectromotry

The association of these techniques with liquid chromatography provides a powerful tool for the identification of components of complex mixtures. After separation by TLC the syringe transfer technique is used because only a small amount (i.e., 20 μg or less) of separated component is required to obtain both its infrared and mass spectra. It is imperative that the minimum amount of eluent is used to effect the transfer and that sample handling be reduced as much as possible so as to eliminate contamination.

To obtain an infrared spectrum, the glass syringe (Fig. 6.14) is filled with a suitable eluent, usually acetone, and the solute eluted through the needle, drop by drop, onto a pile of 10 mg of ground and dried KBr powder. Each drop is allowed to evaporate completely before the addition of the next drop. About 20 drops are usually sufficient to transfer the whole of the adsorbed solute from the silica gel to the KBr. The KBr powder and solute are then mixed with a microspatula and pressed into a 1.5-mm-diameter disk. Excellent spectra are obtainable on 5–30 μg of solute on the Perkin-Elmer 521 infrared spectrophotometer with its microsampling assembly and beam condenser. The presence of this unit increases the path length of the sample beam. This is compensated for by inserting a partially evacuated (500 mm Hg), 1-m gas cell and a beam attenuator in the reference beam.

To obtain a mass spectrum of the separated solute, the desorbed solution is forced through the hypodermic needle and fed directly into the insertion probe of the mass spectrometer.

6.13.5. Flame Ionization Detection

An apparatus has been described by Cotgreave and Lynes[54] in which the sample components separated on a TLC plate are successively pyrolyzed by a slow-moving furnace into a stream of nitrogen which passes into a flame ionization detector. Each zone is recorded on a strip chart as a peak having an area proportional to the amount of material present. The work was carried out with 0.3 mg of solute on 25 × 12 mm plates coated with Silica Gel H. Although moderately successful, several problems remain: (a) extension to conventional-size TLC plates; (b) improvement of resolution; (c) increase of sensitivity; (d) elimination of impurities from adsorbents and solvents; and (e) increase in scanning speed.

In addition it has yet to be established that the procedure can produce quicker and more reliable results than other destructive quantitative techniques such as densitometry.

A similar apparatus has been described by Padley[55]. The separation is carried out on a thin glass rod (0.45 mm in diameter) coated with adsorbent. After development of the chromatogram the components are detected by passing the glass rod through the flame of a flame ionization detector at a speed of 35 cm min^{-1}. The output from the detector is displayed on a strip-chart recorder.

6.14. PRESENT STATUS AND FUTURE OUTLOOK

We have seen in this chapter that TLC is a simple technique utilizing low-cost equipment and materials. It enables not too difficult separations to be quickly carried out and, when combined with intelligent use of revealing reagents and R_F values, provides a powerful analytical method of qualitative functional-group analysis.

In extending the scope of TLC by attempting more sophisticated separations, however, the limitations of the technique become apparent and its performance suffers in comparison with the newly developing high-efficiency, high-speed column chromatography, where experimental variables can be more easily controlled and, eventually, automated.

In quantitative analysis, too, the extra effort required and the cost of ancillary equipment is out of proportion and produces results that can only be regarded as semiquantitative. Quantitative liquid chromatography in the future will almost certainly be carried out more conveniently and accurately in liquid chromatographs in much the same way as gas chromatography is carried out today.

To put TLC into historical perspective, it will be seen to be bridging an important gap between the classical chromatography carried out in short, wide tubes and the new, high-speed, high-performance column chromatography which is described in the next chapter. Nevertheless, the TLC kit will join the standard pieces of equipment–the analytical balance, the pH meter, the infrared spectrometer, and the gas chromatograph–which are to be found in any modern analytical laboratory.

REFERENCES

1. Izmailov, N. A., and Shraiber, H. S., *Farmatsiya* **3**, 1 (1938).
2. Consden, R., Gordon, A. H., and Martin, A. J. P., *Biochem. J.* **38**, 224 (1944).
3. Kirchner, J. G., and Miller, J. M., *Ind. Eng. Chem.,* **44**, 318 (1952).
4. Kirchner, J. G., and Miller, J. M., *J. Agr. Food Chem.* **1**, 512 (1953).
5. Kirchner, J. G., Miller, J. M., and Keller, G. I., *Anal. Chem.* **23**, 420 (1951).
6. Miller, J. M., and Kirchner, J. G., *Anal. Chem.* **24**, 1480 (1952); **20**, 2002 (1954).
7. Stahl, E., *Pharmazie* **11**, 633 (1956).
8. Stahl, E., *Chemiker Z.* **82**, 323, (1958).
9. Pelik, N., Bolliger, H. R., and Mangold, H. K., *Advan. Chrom.* **3**, 85 (1966).
10. Fairbairn, J. W., and Relph, S. J., *J. Chromatog.* **33**, 494 (1968).
11. Peifer, J. J., *Microchim. Acta* 529 (1962).
12. De Zeeuw, R. A., *Anal. Chem.* **40**, 2134 (1968).
13. Korzun, B. P., and Brody, S., *J. Pharm. Sci.* **53**, 454 (1964).
14. Brenner, M., and Niederwieser, A., *Experientia* **17**, 237 (1961).
15. Halpaap, H., *Chem. Ing. Tech.,* **35**, 488 (1963).
16. Killer, F. C. A., and Amos, R., *J. Inst. Pet.,* **52**, 315 (1966).
17. Wieland, T., and Determan, H., *Experientia* **18**, 431 (1962).
18. Niederwieser, A., and Brenner, M., *Experientia* **21**, 50 (1965).
19. De Zeeuw, R. A., *Anal. Chem.* **40**, 915 (1968).
20. Niederwieser, A., and Honegger, C. G., *Advan. Chromatog.* **2**, 123 (1966).
21. Niederwieser, A., *Chromatographia* **2**, 23, 362 (1969).
22. Snyder, L. R., and Saunders, D. L., *J. Chromatog.* **44**, 1 (1969).
23. Stahl, E., *J. Chromatog.* **33**, 273 (1968).
24. Dallas, M. S. J., *J. Chromatog.* **17**, 267 (1965).
25. Giddings, J. C., and Keller, R. A., *J. Chromatog.* **2**, 626 (1959).
26. Brenner, M., and Niederwieser, A., *Experientia* **16**, 378 (1960).
27. Purdy, S. J., and Truter, E. V., *Analyst* **87**, 802, (1962).
28. Goldman, J., and Goodall, R. R., *J. Chromatog.* **32**, 24 (1968); **40**. 345 (1969).
29. Latner, A. L., Molyneux, L., and Dudfield Rose, J., *J. Lab. Clin. Med.* **40**, 345 (1969); **43**, 157 (1954).
30. Dallas, M.S.J., *J. Chromatog.* **33**, 537 (1968).
31. Shellard, E.J., and Alam, H. Z., *J. Chromatog.* **33**, 347 (1968).
32. Amos, R., *J. Inst. Pet.* **54**, 9 (1968).
33. Brain, K. R., and Hardman, R. J., *J. Chromatog.* **38**, 355 (1968).
34. Dobies, R. S., *J. Chromatog.* **40**, 110 (1969).
35. Heathcote, J. G., and Haworth, C., *J. Chromatog.* **43**, 84 (1969).
36. Favretto, L., Pertoldi Marletta, G., and Favretto Gabrielli, L., *J. Chromatog.* **46**, 255 (1970).
37. Jänchen, D., and Pataki, G., *J. Chromatog.* **33**, 391 (1968).
38. Pataki, G., and Niederwieser, A., *J. Chromatog.* **29**, 133 (1967).
39. Zurcher, H., Pataki, G., Borko, J., and Frei, R. W., *J. Chromatog.* **43**, 457 (1969).
40. Sawicki, E., and Johnson, H., *J. Chromatog.* **23**, 142 (1966).
41. Frodyma, M. H., and Frei, R. W., *J. Chem. Ed.* **46**, 522 (1969).
42. Stahl, E., and Jork, H., *Zeiss Information* **68**, 52 (1968).
43. Schulze, P. E., and Wenzel, M., *Angew. Chem. Internat. Edn.,* **1**, 580 (1962).
44. Wilde, P. F., *Thin Layer Chromatography,* Unit Trade Press London p. 29 (1964).
45. Snyder, F., *Anal. Biochem.* **9**, 182 (1964); **11**, 510 (1965).
46. Amos, R., *J. Chromatog.* **46**, 343 (1970).
47. Honegger, C. G., *Helv. Chim. Acta,* **45**, 1409 (1962).
48. Shellard, E. J., and Alam, M. Z., *J. Chromatog.* **32**, 472 (1968).
49. Shellard, E. J., and Alam, M. Z., *J. Chromatog.* **35**, 72 (1968).
50. Graham, R. T. J., Bark, L. S., and Tinsley, D. A., *J. Chromatog.* **39**, 218 (1968).
51. Siggia, S., *Quantitative Organic Analysis,* Wiley, New York (1963).
52. Sawicki, E., Stanley, J. W., Elbert, W. C., and Pfaff, J. D., *Anal. Chem.* **36**, 497 (1964).

53. Tumlinson, J. H., Minyard, J. P., Hedin, P. A., and Thompson, A. C., *J. Chromatog.* **29,** 80 (1967).
54. Cotgreave, T., and Lynes, A., *J. Chromatog.* **30,** 117 (1967).
55. Padley, F. B., *J. Chromatog.* **39,** 37 (1969).
56. Amos, R., *X. Anal. Chem.* **236,** 350 (1968).
57. Stahl, E. (ed.), *Dünnschicht-Chromatographie,* Springer-Verlag, Berlin–Heidelberg–New York, 2nd ed. (1967).
58. Kirchner, J. G., Thin-layer chromatography, in *Technique of Organic Chemistry,* Vol. XII, Perry, E. S., and Weissberger, A., eds., Wiley–Interscience, New York (1967).
59. Mangold, H. K., in Ref. 57.
60. Wood, B. A., in *Quantitative Paper and Thin Layer Chromatography,* Shellard, E. J., ed., Academic Press, London and New York (1968).
61. Kaiser, R., in Ref. 57.

FURTHER READING

Stahl, E. (ed.), *Dünnschicht-Chromatographie,* Springer-Verlag, Berlin–Heidelberg–New York, 2nd ed. (1967) (English edition, 1969); Kirchner, J. G., Thin-layer chromatography, in *Technique of Organic Chemistry,* Vol. XII, Perry, E. S., and Weissberger, A., eds., Wiley–Interscience, New York (1967). (These two books contain major reviews of the techniques and applications of TLC up until 1966.)

Shellard, E. J., (ed.), *Quantitative Paper and Thin Layer Chromatography,* Academic Press, London and New York (1968). (The proceeds of a symposium which reviewed recent developments in various quantitative techniques in paper and thin-layer chromatography.)

Snyder, L. R., R_F values in thin-layer chromatography on alumina and silica, in *Advan. Chromatog.* **4,** 3)1967). (Presents a general theory, which is verified using literature data, for the correlation and prediction of R_F values in TLC.)

3rd International Symposium on Reproducibility in Paper and Thin Layer Chromatography, *J. Chromatog.* **32,** 1–409 (1968). (Original papers and reviews are presented which discuss factors affecting R_F values and various quantitative methods of analysis.)

Chapter 7

Equipment for Column Chromatography

7.1. INTRODUCTION

Until comparatively recently equipment for liquid chromatography was relatively simple and unsophisticated. Apart from glassware one might have used a vibrator to consolidate the bed of stationary phase in the column, perhaps have used a syringe to add sample to the head of the column, and, at the column exit, have collected effluent with a fraction collector. Today the picture is very different and, at the time of writing (late 1970), completely integrated commercial liquid chromatographs are available and cost $2–20M, reflecting the cost of the various newly evolved and often complex components which make up a modern high-performance liquid chromatography system.

Because the evolution of intrumentation is proceding at such a fast pace today, any specific items may rapidly become obsolete and so the emphasis in this chapter will be to discuss the principles on which current equipment is based, its present limitations, and possible future development and improvement. So that in the short term this book may also be a useful reference source, the chapter contains an appendix covering some of the more recently introduced equipment for column chromatography.

In discussing the technique of modern column chromatography we have already noted those aspects of what we might call the "hardware" which distinguish it from the traditional technique. These features can for convenience be grouped under five headings:

1. Pneumatics: the generation and measurement of the relatively high pressures needed for high-speed, efficient liquid chromatography.

2. Gradient elution devices: the generation of eluents of changing characteristic during the development of a chromatogram.

3. Sample injection: the introduction of samples into a liquid chromatograph in such a way that band spreading is minimized.

4. Column construction: the optimum geometry, coupling, and packing of columns and their temperature control.

5. Effluent monitoring: the detection, identification, and determination of separated components of the sample at the column exit.

These topics will be considered in the order given. Two excellent summaries of suppliers of equipment have been published([1, 2]) and the appendix to this chapter seeks to bring those lists up to date.

It should be remembered that in the not too distant future there will not be much incentive to purchase separately these items of "hardware," as complete chromatographs will be available. Nevertheless, an understanding of the components of such instruments will be necessary and the following seeks to provide the required information.

7.2. PNEUMATICS

The permeability of long, thin columns packed with tiny particles of stationary phase is low. To achieve even low volumetric flows (0.01–10 ml min^{-1} is the normal volumetric range of flow) of eluent through such columns, driving pressures well above that provided by the traditional gravity feed are essential. Such necessary driving pressures can be provided either by suitable control and application of the fairly high pressures available from commercial compressed gas cylinders or, alternatively, by using pumps.

7.2.1. Use of Compressed Gas

Compressed gases are generally available at pressures up to in excess of 100 atm, adequate for some of the efficient column chromatography of today. An important feature which attracts one to use a compressed gas is that the pressure is available continuously, as distinct from plunger and diaphragm pumps, which generally operate intermittently. That is to say, the eluent driven by compressed gas flows smoothly at an unvarying rate, whereas pumped eluent moves in a more or less rapid succession of pulses with velocity varying from zero to some maximum value. To damp out these pulses requires extra equipment, while a pump is in addition more expensive than a supply of compressed gas.

It is not sufficient merely to connect a high-pressure gas cylinder to the reservoir of eluent at the head of a chromatographic column. The driving gas would dissolve to some extent in the eluent, from which it might come out of solution during passage through the column. Bubbles of gas or vapor in a column are seriously detrimental to performance, leading to channeling

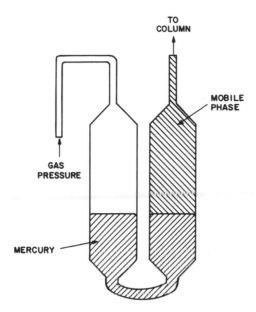

Fig. 7.1. Driver vessel with mercury barrier.

and loss of resolution. Thus dissolution of gas in the mobile phase has to be prevented; this is achieved by separating gas from liquid by a flexible but impermeable barrier. For pressures up to about 5 atm a thick-walled glass vessel containing a mercury interface (see Fig. 7.1) is effective, and simple to construct. For higher pressures metallic construction of these driver-vessels is essential and, since mercury does not amalgamate with steel or iron, vessels of these materials are suitable. An alternative approach is to use a plastic diaphragm. In its simplest form, a form which has been commercialized by at least two manufacturers, the eluent is contained in a polyethylene bottle which is inside a steel pressure vessel. Gas pressure is applied to the space between the metal shell and the plastic, as shown in Fig. 7.2, so that eluent is progressively expelled from the bottle into the chromatographic column.

A limitation of all these gas pressurizing systems is that the total volume of eluent which can be delivered without interruption is restricted to the capacity of the driver vessel and normally is in the range 0.1–1.0 liters. Despite this and the limitation of maximum available pressure noted earlier, there is no doubt that much routine high-efficiency column chromatography could be undertaken cheaply with these simple driving systems.

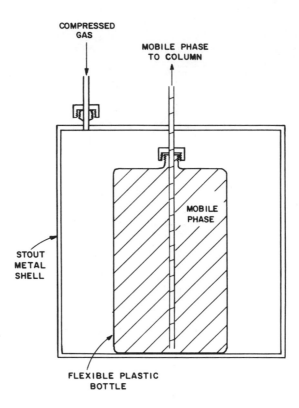

Fig. 7.2. Driver vessel with plastic diaphragm.

7.2.2. Pumps

Although there are many different types of pump in use in laboratory and chemical plant operations, only a few are, as yet, being used for column chromatography. Certain types of pump can only deliver liquid against relatively low back pressures, up to perhaps 5 or 10 atm, and are not therefore an attractive proposition for application to a technique where pressures at least an order of magnitude larger are frequently needed for optimum performance. Rotary pumps, which can deliver liquids essentially pulse-free, are normally unsuitable for delivery of low volumetric flow at high pressures, and peristaltic pumps, which also deliver relatively steady flows of liquid, are limited to use in low-back-pressure systems. Thus the choice of pump for column chromatography, where high performance and high speed are required, usually is restricted to the various positive displacement machines in which the driving force comes from the movement of a plunger, diaphragm, or bellows. Pumps of these types generate a pulsating liquid flow. An advantage of all of the pumping systems over the gas pressurizing approach is that there is no limit to the volume of eluent which can be pumped without interruption, so that pumps are essential for repetitive, automated column chromatography.

In spite of the limited applicability of certain types of pump for our purposes, a short account of each follows, so that the reader may choose for himself that most suited for his requirement.

7.2.3. Peristaltic Pumps

In a typical peristaltic pump the liquid is contained in an elastic tube which is in contact with a curved track. A rotor, carrying several rollers, flattens the tube against the track as it rotates, driving the liquid forward. As the roller passes and the tube returns to its normal shape, more liquid is drawn forward, to be further driven on by the following roller. The advantage of peristaltic pumps briefly are that they are self-priming, relatively pulse-free, can be used with corrosive liquids, and can accommodate several channels in parallel (i.e., several tubes, each pressurizing a separate column, can be actuated by one rotor). To their disadvantage of restricted driving pressure (10 atm is a realistic maximum) must then be added the possibility of eluent contamination by plasticizer from the elastic tube and the fact that some organic solvents are not compatible with plastic tubing. The cost of pumps of this type ranges up to about $500.

7.2.4. Reciprocating Pumps

These pumps, including plunger, diaphragm, and bellows pumps, operate through the application of a reciprocating motion of the driver (plunger, diaphragm, or bellows) which in turn forces the liquid forward in a series of pulses.

The principal features which characterize pumps of this class are set out in Table 7.1.

Table 7.1. Characteristics of Reciprocating Pumps

	Plunger	Bellows	Diaphragm
Operating pressure, atm	1–100	0.5–25	1–3000
	Generates pulsating flow	Generates pulsating flow	Generates pulsating flow
	Time-averaged flow rate not very constant at low flow rates	—	Time-averaged flow rate very constant at low flow rates

From Table 7.1 it will be seen that bellows pumps are not particularly suited for column chromatographic work because of their relatively low operating pressure. Of the other two types, the diaphragm pump has two distinct advantages, the higher pressure range and the valuable "constancy" of flow at low flow rates. However, it must be remembered that the basic flow from any reciprocating pump is pulsating, and generally must be damped. The difference in undamped flow profile from plunger and diaphragm pumps is exaggerated in Fig. 7.3. Thus for maximum flexibility and optimum performance under conditions consistent with high speed, high-efficiency column chromatography, diaphragm pumps are the most suitable of the types presently available.

7.2.5. Rotary Pumps

The basic action of a rotary pump is to trap liquid between the teeth or vanes of a pair of rotors and thus drive it forward. Essentially, such pumps are relatively pulse-free, but suffer from the disadvantage that it is very difficult to obtain low flow rates (i.e., those required for column chromatography) from them even with viscous liquids, let alone the low-viscosity mobile phases used for chromatographic work.

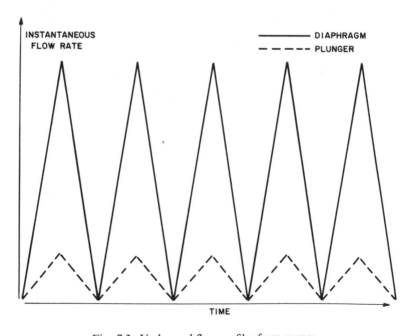

Fig. 7.3. Undamped flow profiles from pumps.

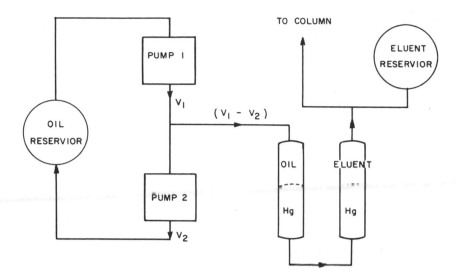

Fig. 7.4. Differential rotary pump system [after Johnson and Cantow(3)].

An ingenious system to overcome these difficulties and provide a constant (low) volume, pulseless flow for column chromatography has been described by Johnson and Cantow(3). They operated two constant-volume, gear-type pumps differentially to obtain the desired low flow rate to the column and, to get the necessary volumetric constancy, operated the pumps on a high-viscosity mineral oil. The expelled oil is used to drive the required eluent through the column, being separated from it by a mercury layer; this part of the system thus resembles that described earlier for converting gas pressurization, except that the pressures generated are higher and the whole equipment is constructed of steel. A simplified schematic of Johnson and Cantow's system is shown in Fig. 7.4.

7.2.6. Constant-Drive Piston Pumps

Probably the simplest pumping system consists of a piston driven at constant rate down a cylindrical reservoir containing the mobile phase. A completely pulse-free stream of liquid is thus forced through the column at a variable, predetermined rate.

The total volume of mobile phase is limited to the capacity of the cylindrical reservoir. This in turn is limited by size and weight considerations and also by the maximum cross-sectional area which can be tolerated. If this becomes too great, there may be difficulties in producing a low enough forward rate of movement of the piston to generate the relatively low flow rates often required. Another size-limiting factor is the difficulty of accurately machining large pistons and cylinders to the necessary close tolerances. At present these pumps are available for LC with capacities not exceeding 1–2 liters.

It is now practicable to add a second component which enables gradient elution to be used with this system.

7.2.7. Methods of Damping Pulsating Flows

For optimized chromatography the linear velocity of the mobile phase should remain constant, or in certain circumstances be progressively changed. The reciprocating pumps and, to a much lesser extent, the peristaltic pumps generate pulsating flows such that the mobile phase is running at its optimum velocity for only a small proportion of the time. Means of damping out the velocity fluctuations are necessary and several approaches are available.

Methods for pulse damping include the use of differentially operating pumps, Bourdon tubes, "piston" dampers, and hydropneumatic accumulators.

Use of a series of pumps operating out of phase can produce a reasonably constant flow, at a relatively high cost. Various positive displacement pumps are available in "multiple-head" versions (that is, a set of individual

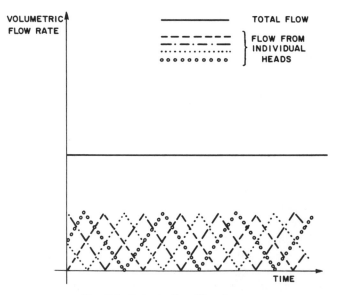

Fig. 7.5. Output of multihead pumps.

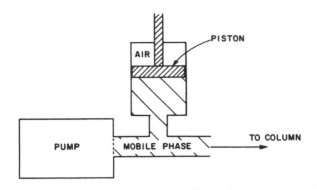

Fig. 7.6. Piston damping system.

delivery heads of equal capacity arranged in parallel and all driven by one pump).

The net delivery from such a system with four heads operating 90° out of phase is illustrated in Fig. 7.5.

Bourdon tubes are flexible, coiled, small-bore metal tubes closed at one end and filled with fluid. When the fluid is pressurized the tube tends to uncoil, dissipating some of the energy of the pressurized fluid and redistributing it during the later part of the pump cycle. Thus, when a Bourdon tube is connected, or "teed," into a pump–column system the energy redistribution can be sufficient to achieve substantial damping of the flow. Variations on this energy redistribution theme are available in which other elastic media are employed, for example, rubber tubing or air pockets.

Piston dampers use the medium of an air pocket to absorb and redistribute pump energy. A schematic of such a damper is shown in Fig. 7.6. As the pump displaces liquid, the air is compressed and a proportion of the displaced liquid is accommodated within the cylinder. As the pump delivery falls to zero, the air expands, expelling liquid forward into the column, thus achieving overall a measure of flow constancy. The principal problem with this type of damper is the careful machining of the piston and cylinder required for successful operation at high pressures.

Hydrodynamic accumulators are similar in design to the gas pressurizing systems described earlier. A typical example shown in Fig. 7.7 consists of an elastic bag in a steel shell. The bag is filled with gas—normally air—and the space between it and the shell is filled with mobile liquid phase being pumped. Compression of the enclosed gas and its subsequent re-expansion during the pumping cycle provide the necessary flow damping. Devices of this type are claimed to be operable to 400 atm and their principal limitation arises from the probability of attack by the mobile phase on the elastic material of the

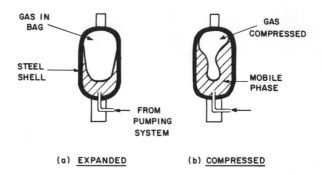

Fig. 7.7. A hydrodynamic accumulator.

bag. This restricts the range of eluents which can be used. In addition additives in the polymeric material from which the bag is made may be leached out by the solvents and can lead to serious contamination problems. However, for "mild" solvents the cheapness of these devices ($\sim$\$50) makes them an attractive proposition.

7.2.8. Solvent Degassing

It is desirable to remove dissolved air from the mobile phase before it enters the columns for two reasons: (1) Air bubbles may form in detector cells and interfere with their operation. This is particularly important when a differential refractometer is used, since it is usually maintained at a temperature slightly above that of the columns. (2) Oxidation may occur in the columns.

Removal of air is usually accomplished by heating the mobile phase to just below its boiling point before it enters the pump.

7.3. GAUGES

Of all the accessories for the liquid chromatographer pressure gauges are probably the most dispensable. They are used to show the column inlet pressure which is related, via column permeability, to the important property, linear velocity of the eluent. Apart from giving a measure of column permeability (eluent velocity is calculated more conveniently from measurements of volumetric flow and column dead volume), knowledge of the inlet pressure is useful as a safety feature. However, trip switches which deactivate the pump if the pressure exceeds a predetermined level are better as safety features since they function automatically.

A wide variety of gauges are available commercially, with ranges of from 0–10 atm to 0–2000 atm. They are based on the Bourdon tube principle.

7.4. GRADIENT ELUTION SUPPLY

Gradient elution is a most valuable technique in liquid chromatography. In essence the term refers to a process in which the composition of the eluent changes with time. The purpose of so doing is to use a mobile phase of progressively stronger eluting power so that more strongly retained solutes can nevertheless be eluted in reasonable times. Gradient elution is in many ways the liquid chromatographic analog of temperature programming in gas chromatography. The change in mobile phase composition is normally continuous and more or less linear; however, in some circumstances abrupt changes in composition may be helpful.

A variety of simple devices are available for generating eluents of continuously changing composition and which can easily be fabricated in the laboratory. Typical of these devices is that shown in Fig. 7.8, which is composed wholly of glass; the eluents A and B do not have to be pure compounds. For example, a very slowly increasing composition of, say, toluene in heptane could be generated by having eluent B as pure heptane and eluent A a few per cent of toluene in heptane. Extensions of this type of system are possible with multiple-eluent reservoirs linked together. Essentially linear composition gradients are produced, at rather low operating pressures (up to about 10 atm).

A more complex unit for generating an optimized (logarithmic) solvent program designed for operation at up to 400 atm has been described by Snyder and Saunders[4]. In this system the eluent is charged to a series of "holding" columns: each column is packed with glass beads to minimize

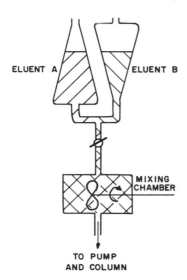

ELUENT A

ELUENT B

MIXING CHAMBER

TO PUMP AND COLUMN

Fig. 7.8. Glass equipment for generating linear elution gradients.

intermixing of the contents of each column as they are serially forced into and through the separating column. Each holding column is filled with a chosen eluent in sequence of increasing eluotropic strength so that a discontinuous gradient is generated. Because there is some mixing between the contents of adjacent holding columns, the actual changes in the solvent are not absolutely sharp. This system has the advantage that very precise eluent compositions can be preprepared and these are not very seriously degraded by the sub-sequent mixing; in addition, because the eluent sequence is held in steel columns, very high pressures can be used. On the other hand, the capacity of the system tends to be limited (Snyder's limit was 330 ml) and the need for multiple holding columns makes the approach cumbersome. It is probably only of value where rather special solvent programs involving mixtures of a large number (e.g., more than five) of pure materials are required to achieve particularly difficult separations. Other systems have been described for generating eluent gradients based on a holding column in which the desired eluents are held separate by density differences so that the filled column contains discrete layers of progressively changing eluting power[5]. This approach has been used in the adsorption chromatographic separation of heavy petroleum fractions on alumina[6].

A few commercial variable-gradient generators are available; normally these are based on the use of two pumps which can be programmed through a range of flows and operations so that the mixed total output from the two can be made to follow fairly elaborate gradients of pH, eluotropic strength, density, or similar properties. These devices at present are limited to outputs of about 100 atm but, of course, one can use ancillary high-pressure pumps to boost this to any desired level.

The authors believe that at present the most economical and con-venient system for all but the most complex gradients is a combination of a mixing system like that of Fig. 7.8 with a high-pressure pump to deliver the eluent to the column.

7.5. SAMPLE INJECTORS

Samples for separation by LC are injected as liquids, in a suitable solvent if they are originally solids. Injection is either by syringe or a sampling valve, as in gas chromatography.

Syringe injection is cheap, convenient, and flexible in that any volume from fractional microliters upwards can be dispensed easily. It has several basic limitations which severely restrict its use in high-performance liquid chromatography. The crux of the problem is that these injections normally involve use of a self-sealing polymeric septum, through which the syringe needle is forced so that the sample can be introduced into the mobile phase.

These septa are incapable of leak-free operation at pressures above about 20 atm; indeed, they may even be extruded back through the retaining metal nut at such pressures. In addition organic additives in the polymer—extender oils and antioxidants, for example—and low polymers or unreacted monomer may well be leached out of septa by the mobile phase and foul the column, detector, and any collected fractions. These deficiencies can be minimized. The system described by Hurrell and Amos[7] for injection into high-pressure gas chromatographs is a useful expedient to extend the usable pressure range of septa; in essence it requires that the septa be retained by a nut with an orifice just big enough to allow passage of the needle, thus giving maximum rigid support to a pressured septum. Further support, together with a degree of protection from the leaching properties of the eluent, may be obtained by wrapping a septum in polytetrafluoroethylene tape as has been described by Scott et al.[8]. By these devices septa may be used and syringe injection accomplished against pressures of up to about 100 atm. Under these conditions quantitative transfer of sample from syringe to column cannot be expected.

Samples may also be introduced onto columns from a syringe by using a demountable injection port; when an injection is to be made mobile phase flow is stopped, the port opened, and the sample introduced. The port is then closed after ensuring that no air has been trapped, and eluent flow resumed. This is a slow and inelegant method which, because the eluent flow takes some time to build up to its set level after the discontinuity, makes for difficulty in identification by retention measurements and also renders the assessment of column performance with respect to eluent velocity imprecise. Nevertheless, it is a practical expedient which can be used in work at exceedingly high pressures if necessary.

The use of sampling valves enables some of the problems already mentioned which are associated with syringe injection to be avoided. A typical liquid sampling valve and mode of operation are shown schematically in Fig. 7.9. The valve is of the sliding variety, typically using all polytetrafluoroethylene construction. The sample fills a hole of known capacity machined through the barrel of the valve. The barrel is then pushed to a position where the sample volume is swept by the eluent onto the column. Pneumatic actuation of such valves is possible and this permits automatic sampling of streams for routine product control. Rotary valves are also available which can be obtained in various materials, although they are normally of stainless steel construction.

Although the range of sampling valves available is growing quite rapidly at the time of writing, they all suffer from three limitations which need to be overcome before they are adopted for widespread use. These limitations are, in order of increasing seriousness, the limited range of sample volumes

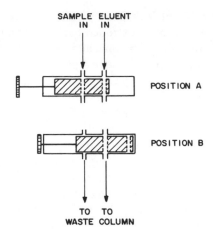

POSITION A   SAMPLE CHARGED TO VALVE
POSITION B   SAMPLE SWEPT ONTO COLUMN

Fig. 7.9. Operation of a slide sampling valve.

which may be injected with any one valve, the limited maximum temperature to which the valve can be taken, and the limited maximum pressure at which it can be operated. Looking at these limitations in turn, that of restricted sample size is no more than a modestly expensive inconvenience. Inconvenient because, at best, the loop or cavity (in which the sample is trapped and then carried into the mobile phase path) has to be changed since each one is inevitably of a fixed volume. Expensive because, to cover the range of sample volumes required, it may be necessary to buy several different valves. This inflexibility of valves puts them at a disadvantage compared with syringes. The temperature limit of most valves is around 150–200°C and is set by the deformation of moving parts or seals, or even by degradation of some polymeric components. In liquid chromatography the temperature limitation primarily acts through restricting application so that very insoluble materials (which only have a sufficient solubility at perhaps over 200°C in an otherwise ideal solvent) or very viscous liquids cannot be handled. Pressure limitations of valves are by far their most serious disadvantage, although it is in this respect that progress is most encouraging. Valves said to be capable of operation at up to 200 atm are now available. To put this figure into perspective, the pressure drop across a 1 m column length, 4 mm i.d., packed with stationary phase of about 40 μm average particle diameter is about 10 atm for a flow rate of 1 ml min^{-1}. Thus an inlet pressure of 200 atm is more than adequate for operation of extremely long columns at flow rates well above the optimum for maximum resolution. In other words the pressure limitations of the better sampling valves are not at present serious; it should however be borne in mind (a) that most sampling valves now available are not capable of operation above 50–100 atm and (b) it appears likely that stationary phases with particle diameters in the range 1–10 μm will prove to be much more satisfactory than the coarser ones now being used—columns packed with the very fine particles will be much less permeable than the one exemplified above (an order of magnitude decrease in mean particle diameter requires a corresponding increase in pressure; thus with 4-μm particles an inlet pressure of 200 atm would be adequate only for columns up to about 5 m long).

Sample injection devices—syringes or valves—are thus seen to be just adequate for the state of column chromatographic art today. Valves are the only convenient practicable method for injection against very high pressures and encouraging improvements in valve design and fabrication are continually being commercialized. The prospect that injection devices will improve in step with other features of column technology look promising.

7.6. COLUMNS AND FITTINGS

Old-fashioned, inefficient column chromatography almost invariably involved the use of glass columns. Because of the high pressures used in the "new" chromatography, normally in excess of 15 atm, glass is of limited value. The maximum operating pressure for even thick-walled glass tubing is about 100 atm. Plastic columns are even more limited, so that, for practical purposes, only thick-walled glass or metal columns can be considered. Stainless steel, copper, and aluminum are the obvious choices, as in gas chromatography.

Lengths and diameters of columns affect the degree of resolution obtainable and their permeability. At the same time these dimensions are important because of their controlling influence on methods of column packing, for example, packing very-small-diameter columns (2 mm i.d.) to achieve good consolidation of the packed bed is difficult. Similarly, very long columns almost inevitably have to be bent or coiled in some way for convenience. This normally has to be done before the packing operation since the action of manipulating packed columns may lead to a serious decrease in efficiency. Very few studies on column packing procedures have been published, but the best so far appear to be ones involving consolidation of an adsorbent by tamping[9–11]. Tamping can only be done in straight columns, so that the maximum length which can be conveniently packed is limited to 1 m, or perhaps a maximum of 2 m. If simple methods of column packing could be devised which were capable of generating HETP's less than 1 mm after coiling the packed column or by packing the empty coil, then there is no reason why rather long lengths of easily workable copper tubing should not be used, providing the samples to be handled did not interact with the metal.

As a guide, the following procedure for packing straight metal columns up to 1 m in length is offered:

1. Fix an appropriate connector on the lower end of the column, using this connector to support a suitable porous plug which will retain the column packing (small pore filters of the Millipore and Gelman types are suitable).

2. Pour into the column sufficient dry packing to occupy about 2 cm length before consolidation.

3. Rest a metal or Teflon-ended plunger on the top of the packing bed (the plunger end should be a close fit for the column, just free enough to move without undue effort).

4. Gently tap the bottom of the column on the floor so that the plunger, or "tamping" rod, compacts the previously loose bed of stationary phase.

5. During the tamping operation a few gentle taps on the walls of the column can be made to further assist in consolidating the packing bed.

6. When consolidation of the first quantity of packing is complete (no further compacting observable) repeat with a further 2 cm loose bed and so on until the column is completely full.

For a 1 m length of column the above operations take about 15 min, after familiarization with the technique. This procedure is recommended for stationary phase particle sizes above about 20 μm. It is not suitable for finer materials and there is as yet no one method for dealing with such materials which could be recommended with any confidence.

Column fittings needed are the injection port (unless a sample valve is used), intercolumn connectors, and precolumn and postcolumn connections. Critical factors are the dead volumes of all of these fittings through which the sample passes; in these cases this volume should be a minimum. The length of connectors should always be the smallest possible, but the optimum diameter may not be the practical minimum because, depending on flow rate, with very small diameters, flow may pass from streamline to turbulent and some resolution would be lost as a consequence. Thus it is necessary to find the optimum diameter, as a function of flow rate, by experiment. A suitable design of column connector is described in Ref. 9. The same paper describes a sample injection port carefully designed to minimize band spreading. Essentially, it consists of an axially mounted, narrow-bore needle guide down which the syringe needle is pushed into the top of a bed of small-diameter (~ 50 μm) glass beads. The glass bead bed, about 8–10 cm long, serves to diffuse the introduced sample more uniformly over the whole cross section before reaching the actual separating bed of packing.

Not all workers use this glass-bead zone at the column head; in practice, with syringe injection, blockage of the needle can only too easily occur. Further, while Scott et al.([8]) and Stewart et al.([9]) have found the glass-bead bed advantageous, others([10]) believe that inevitable diffusion in the bed contributes to the solute band spreading and prefer to inject into a short (~5 mm) bed of Teflon or glass fiber.

Columns are still normally operated at ambient temperature. However, there are several very good reasons why column temperature should be controlled and variable:

(a) Solutes may be insufficiently soluble at low temperatures.
(b) Mobile phases may be too viscous at ambient temperature (hard to pump).
(c) Temperature has an influence on relative retentions such that a particular separation may be facilitated by change of temperature.
(d) In liquid–liquid partition close temperature control is essential to maintain a fixed phase-pair equilibrium and avoid column "bleed."
(e) Increase in temperature decreases the mobile phase viscosity; this decreases peak broadening.

Thus the liquid chromatographs becoming available, and their successors, will resemble gas chromatographs in having their columns in a temperature-controlled oven. As in GC it will certainly be necessary to have very close control of temperature, i.e., to $\pm 0.1°C$ or better, for use of precise retention data for the more difficult identification problems and also to ensure stability of the more critical partition systems. For these cases liquid thermostats may well be used. The possibility of temperature programming LC columns has been suggested([12]), but some authorities prefer gradient elution as a much more effective means of attaining the same end([13]).

7.7. GRINDERS AND SIZING EQUIPMENT

There is ample theoretical([14]) and experimental([9, 15, 16]) evidence that very small particles, of uniform (i.e., narrow range) diameter provide the ultimate in column performance. At present such materials are just becoming commercially available. Taking the most commonly used adsorbent, silica gel, most suppliers still only offer relatively wide-range, coarse particle sizes, i.e., 75 μm upwards. In addition silica gels for thin-layer chromatography are available which, typically, span the 5–50-μm range. To obtain the small, narrow-range, sizes which appear to be desirable (i.e., 5–10 μm, 20–30 μm, etc.) one may have to grind coarser gels and will certainly need to sieve or otherwise size-grade either the ground gel or the TLC grade to the range desired. Thus at present, though hopefully not in the future, the liquid chromatographer needs to be equipped with grinding and size-grading apparatus.

While one can use a pestle and mortar to grind materials manually, this is so time-consuming that a power-operated grinder is really an economic necessity, despite its cost. At least one such unit is commercially available.

There are two standard laboratory approaches to the preparative size-grading of small particles: sieving and sedimentation. Sieves are of two basic types, wire mesh or electroetched metal, and can be operated either dry or with a liquid (usually water) or air flow to aid the sieving operation. Wire mesh sieves are widely used, but frequently are not available for sizina in the range below about 50 μm, though there are exceptions. The minimum screen opening of these sieves is about 25 μm. A disadvantage of wire screen sieves is that the screen may become physically distorted so that some openings may become substantially larger or smaller than their nominal size.

Electroformed nickel sieves are available with openings down to 5 μm. Essentially, these are metal membranes in which holes of the appropriate size have been etched out. Consequently, these sieves cannot be distorted other than by violent ill-usage (though of course they should not be abused). Unfortunately, the total open area of such a sieve is very much less than that of a wire sieve. In consequence, the throughput of sized materials is much lower with the otherwise excellent electroformed sieves.

Sieving dry materials has the self-defeating disadvantage that the build-up of static electricity by interparticulate friction leads to formation of unsievable agglomerates. If a constant stream of water flows through the stack of sieves in use, this effect is eliminated and a more rapid and complete sizing can be achieved. Sieving as a means of obtaining size-graded small particles is a slow task, unsuited to large-scale production, for which sedimentation fractionation appears to be much more satisfactory.

In a typical batch-sedimentation fractionation the wide-range particulate sample is added to a fluid, usually water, contained in a vertical column (see Fig. 7.10). Up-flow of water carries the fine particles away, while the heaviest fall to the bottom and are periodically drained off. The size range of the particles which remain in suspension is principally controlled by the water flow rate and this is normally adjusted by trial and error until the appropriate size range is obtained.

7.8. FRACTION COLLECTORS

For many years column chromatography has been carried out as a separation process to obtain materials for further examination rather than as a fully integrated analytical procedure. The automation of fraction collection has thus been very fully developed. Fraction collectors of a variety of ge-

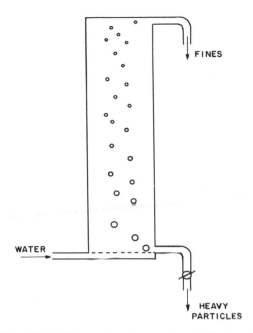

Fig. 7.10. Sedimentation of supports and adsorbents.

ometries, capable of recovering quite small (1 ml) or large (100 ml) fractions and which can, for example, be refrigerated, are readily available with a galaxy of drop counters, timers, etc. to control the collection of desired portions of the eluate. Very many companies market equipment of these types and the reader is referred to recent tabulations of suppliers[1,2].

The newly developing high-performance chromatography, while being essentially an analytical method, nevertheless can usefully provide small samples for further characterization. Although so far as is known no such device is currently available, one can anticipate that small fraction collectors (i.e., much smaller overall dimensions than the current models and designed to collect 0.01–1.0 ml fractions) activated by the detector output, rather than by drop number or time, will soon become available. These devices will resemble fraction collection systems designed for preparative-scale gas chromatography except that the need for freeze-out facilities will not exist.

7.9. DETECTORS

7.9.1. Introduction

Allowing that much of the recent resurgence of interest in liquid chromatography in columns has stemmed from the rather belated recognition that the methods, techniques, and principles so successful with gas chromatography are applicable to liquid phase separations, the single crucial factor that has catalyzed much of the recent activity has been the introduction of continuous detectors. No amount of sophistication of the column and its packing could have compensated for the tedium of fraction collection and examination which otherwise would have been the only way of monitoring column separations. The immediate visual display of the progress of a separation on a strip-chart recorder has enabled much creative effort to be diverted to the more profitable improvement of column design and the screening of potential applications. It is therefore right and proper that these detection systems are described in some detail; none of them is by any means as versatile as the liquid chromatographer would wish–there is no real equivalent either to the katharometer or the flame ionization detector of GC. Therefore much development is going on in this field and significant advances can be expected. In describing the principles on which successful detectors are based we will concentrate on the potential for future development as well as the state of the art as it exists at the time of writing.

Table 7.2 lists properties which an ideal detector would have. It is unlikely that any single detector will have all these desirable attributes; certainly at this time none of those available begins to match up to these ideals. Why are these properties considered necessary?

The desirability of a detector responding to all separated solutes is obvious; it is generally necessary to get a complete picture of the separation achieved and, in the absence of a single, universally responsive device, a series of less universal detectors is necessary which together give the complete elution profile; apart from any other disadvantage such as arrangement, these could be extremely costly to install. As in gas chromatography, specific detectors have a role to play, too; however, reviewing the progress of GC, it will be noted that the detectors which were used almost exclusively in that first decade when frequent and impressive advances were being achieved were virtually universal in their response. Only now, nearly 20 years after the birth

Table 7.2. Properties of the Ideal Detector

Responds to all solutes or else shows predictable specificity
Does not respond to the mobile phase
Has high and predictable response
Is unaffected by changes in ambient temperature and in eluent flow
  rate
Does not contribute to solute band spreading
Provides information on the identity of a detected solute
Is easy to use and cheap
Has a response which increases linearly with solute quantity
Is nondestructive of the solute

of GLC, are specific detectors playing any real part in routine GC. Thus it is concluded that LC has a need for the equivalent of a katharometer or gas density balance.

A major reason why ideally our LC detector will not respond to the mobile phase follows from the fact that in elution chromatography the solute is present at very low concentrations in the column effluent, seldom exceeding 1 %, and often several orders of magnitude more dilute. Response to the enormous excess of eluent requires that its composition and velocity remain very constant, otherwise the "noise" generated by variations in these factors could well swamp any response due to the presence of an eluted solute. Gradient elution is a desirable technique for many separations. The changing composition of the mobile phase may lead to serious, even insurmountable, problems if the detector responds to these changes. To a great extent, therefore, the first two listed attributes of an ideal detector appear to be irreconcilable, although in fact we will see later that there is at least one system of almost universal applicability, and several of more limited scope, which do overcome the problem of response to the eluent.

The stated requirement for high and predictable response follows from the view that in the future column LC will provide high resolution and accurate quantitative analysis of complex samples. As a rule it may be stated that as sample size decreases, the resolution of separated components increases. In high-efficiency LC sample sizes in the range 10^{-2} to 10^{-6} g are used, individual components being present at, say, 1 % of these quantities; hence the need for a system which will detect as little as 10^{-8} g (preferably less) of a solute eluting in about 1 ml of eluent. This need for high sensitivity needs to be stressed; lack of sensitivity is still a significant limitation of many detectors. The maximum concentration of solute in such a peak is at the ppm level. Predictability of response (as a function of molecular structure) is not essential; it is usually possible to prepare calibration curves through the use of data obtained on pure components. However, an appreciation of the molecular features which determine the extent to which a detector responds to a solute is of great value either in estimating a component which is unavailable for calibration or, on occasion, in predicting the type of compound from the magnitude of the response of two detectors to it.

Absolute constancy of flow and temperature are virtually unattainable and the less a detector is sensitive to variations in these factors the easier and cheaper it is to use. For example, rather elaborate measures are necessary to damp out the pulsations in mobile phase flow generated by the pumps needed for high-pressure LC; similarly, close control of detector temperature (which has a profound effect on the magnitude of some physical properties which otherwise would provide an ideal indicator of the presence of solutes in an eluent) requires thermostatting arrangements which, unless extremely care-

Table 7.3. Detectors for High-Resolution Liquid Chromatography

Available commercially; use well documented:
- Refractive index
- Heat of adsorption
- Spectrophotometric
- Transport-ionization
- Radioactivity
- Electrical conductibity

Essentially home-made; relatively little published work:
- Polarographic
  - Heat of reaction
  - Density
- Dielectric constant
  - Vapor pressure

fully designed, may sometimes contribute significantly to band spreading in the postcolumn connections to the detector. Examples of detection systems where these environmental problems are significant will be discussed later.

The need to avoid any unnecessary contribution to solute band spreading is obvious in any analytical situation where high resolution is necessary. There is no point in carefully preparing an excellent column and optimizing its operating parameters if extracolumn factors cause a peak broadening several times greater than the column itself–and this situation can only too easily occur.

In Table 7.2 we have stated that our ideal detector should give qualitative information about the detected solutes. Liquid chromatography is applied to less volatile, and in general more complex, molecules than gas chromatography. Identification through retention measurements is therefore rather more difficult and time-consuming. Hence any information which can be deduced from the detector is helpful, for example, the recognition that specific functional groups or elements are present. Where identification is necessary and difficult, fraction collection for subsequent spectroscopic and chemical analysis becomes vital; in these cases it is necessary to have a non-destructive detector.

In quantitative chromatographic analysis a linear relationship between solute quantity and detector response minimizes the need for calibration. Generally the range of solute quantity over which linear response is obtained varies among the different detectors; other factors being equal, the detector with the widest linear range is preferred.

In commenting on desirable properties of a detector ease of use and cheapness have been cited directly or indireclty in several cases. A guiding rule in analysis should be that the cheapest, simplest, easiest-to-use techniques and methods capable of providing the requisite information should be employed. Liquid chromatography is no exception and the enormous popularity of thin-layer, as opposed to column techniques testifies to the widespread recognition of this "rule."

Against this idealistic background we will now consider detection systems, or principles, which are or may soon become useful for monitoring high-speed, high-efficiency column chromatographic separations. In Table 7.3 these systems are listed, with an indication of the extent to which each is at present being used for this purpose.

7.9.2. Commercialized Detection Systems

(a) Refractive Index Detectors

One of the first detectors with adequate sensitivity and universality of response to become commercially available was the Waters R4 differential refractometer; it formed the brain of the first "modern" liquid chromatograph, the Waters Gel Permeation Chromatograph. Therefore refractometry has played a special role in the growth of column chromatography. Improvements to the original design have been made and other companies have produced refractometers suitable for use in this field and together these various competitive instruments are probably more widely used than any other single type of detector.

Essentially, they provide a continuous record of changes in the refractive index of eluate leaving the end of the column. A schematic diagram of a typical differential refractometer is shown in Fig. 7.11. Light passes through the sample and reference cells and is reflected back through them onto a beam-splitting mirror so arranged that when the liquids in the two cells have the same refractive index the intensity of each beam is the same. However, when the refractive index of one liquid changes, the position of the light beam incident on the splitter changes and the intensities of the split beams no longer remain equal. The differential output of photocells monitoring split-beam intensities is normally displayed on a strip-chart recorder. The peaks

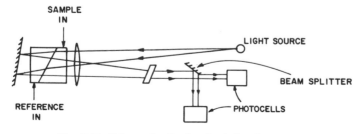

Fig. 7.11. Schematic of refractive index detector.

recorded may be either positive or negative, depending on the refractive index of solutes eluting in the mobile phase, and the size of peaks depends on the amount of solute and on the difference between its refractive index and that of the eluent.

These are several advantages of refractometric monitoring. Probably the most important is the universality of response; almost invariably solute and eluent will differ in refractive index (difference ΔRI) so that almost any substance can, in principle, be detected. The limit of detection is set as noted implicitly above, by ΔRI and the amount of solute. Commercial differential refractometers are claimed to be capable of detecting ΔRI's as small as 10^{-7} refractive index units. As a guide to sensitivity under normal conditions a solute differing by about 0.1 RI units from the mobile phase can be readily detected at a concentration of 1 μg ml^{-1}. Sensitivity can be maximized by choosing an eluent with a very high or very low RI in some circumstances. This is particularly true in gel permeation chromatography, where the mobile phase composition is relatively unimportant compared with the porosity characteristics of the stationary phase. In adsorption and partition work the mobile phase constitution plays a vital part in the separation, so that choice is restricted. A list of refractive indices of organic liquids commonly used in liquid chromatography is given in Table 7.4, see also Table 3.4. It will be noted that both strongly polar and slightly polar eluents can be found near both extremes of the refractive index scale; thus, depending

Table 7.4. Refractive Indices of Common Mobile Phases

Methanol	1.33	Diethylene glycol	1.45
Methyl formate	1.34	Chloroform	1.45
Acetonitrile	1.34	Cyclohexanone	1.45
Diethyl ether	1.35	Carbon tetrachloride	1.46
n-Pentane	1.36	Cyclohexanol	1.46
Acetone	1.36	Trichlorethylene	1.48
Ethanol	1.36	Toluene	1.50
Methyl acetate	1.36	Benzene	1.50
Ethyl acetate	1.37	Ethyl benzoate	1.51
n-Hexane	1.38	Pyridine	1.51
i-Propanol	1.38	Anisole	1.52
Nitromethane	1.38	Chlorobenzene	1.53
n-Heptane	1.39	Furfural	1.53
n-Butylamine	1.40	o-Dichlorobenzene	1.55
Tetrahydrofuran	1.41	Nitrobenzene	1.55
Furan	1.42	Carbon disulfide	1.63
Dioxan	1.42		
Dichloromethane	1.43		
Cyclohexane	1.43		
Dimethylformamide	1.43		

Table 7.5. Solvent Pairs for Gradient Elution with Differential Refractometry

Nonpolar component	RI	Polar component	RI
n-Heptane	1.3855	n-Propanol	1.3854
n-Hexane	1.3754	iso-Propanol	1.3776
n-Propyl ether	1.3807	Methyl ethyl ketone	1.3807

on the nature of the sample, chromatographers have some flexibility toward maximizing sensitivity of detection with adequate elution programs.

Refractometers are nondestructive of sample and, at the low solute concentration met in practice, their response increases linearly with solute quantity.

The disadvantages of refractometric column monitoring are important, though they should not be overemphasized. Despite the inherently high sensitivity of these devices to changes in RI, their ability to detect small quantities of many solutes is scarcely adequate for high-resolution work, with the requisite small samples. A high degree of temperature constancy is required; the temperature coefficient of RI for many organic substances is of the order 10^{-4} units per degree Celsius. Thus to achieve the potential sensitivity of commercial refractometers temperature control to $\pm 0.001°C$ is desirable. Since temperature changes tend to reflect variations in environmental conditions, they are relatively slow processes compared with the rate at which peaks elute from a chromatograph. It seems reasonable, therefore, to expect that electronic compensation for temperature drifting could be incorporated into differential refractometers in the future so that the experimental difficulties this factor currently presents could be substantially diminished. One of the problems arises from the need to precisely thermostat the cell and the liquids flowing into it; in some of the commercial instruments available today the liquid streams pass through quite long tubes within a thermostatted metal block in order to achieve constancy of temperature. While this successfully achieves its aim, it ignores the solute band spreading which inevitably occurs in the long tubes. In high-resolution chromatography (i.e., with columns having HETP < 0.5 mm) this seriously limits the overall performance of the system, as has been shown by Huber[17] and Billmeyer et al.[18]. As a result it is becoming common practice to modify manufacturers' flow systems to eliminate as much as possible of the tubing and take other measures to control environmental conditions and thus minimize temperature variations.

Use of gradient elution techniques with differential refractometers is difficult because of the problem of maintaining the necessarily close equivalence of refractivity of eluent and reference flows. The best approach to minimizing these difficulties which has so far been described is that of Bombaugh and coworkers[19], who have developed useful gradients in which the component pairs have approximately the same refractive index (see Table 7.5).

In summary, differential refractometers are good general-purpose detectors of moderate sensitivity. They are the nearest approach at the time of writing to a universal detector for LC, but will inevitably be superseded by more sensitive and more flexible systems in the future; they are a stopgap.

(b) Heat-of-Adsorption Detectors

The heart of these detectors is a sensitive temperature sensor (normally a thermistor bead) embedded in a material onto or into which eluted solutes are adsorbed. Sorption is exothermic and the heat released leads to a temperature rise measured by the thermistor. Since a chromatographic process depends on the interaction of solutes with a retaining medium, it follows that by locating the thermistor in the same material as is contained in the column all solutes (other than those unretained by the column) will in principle be detected. Further, since long retention is due to a high degree of solute–stationary phase interaction, late eluting components (with correspondingly

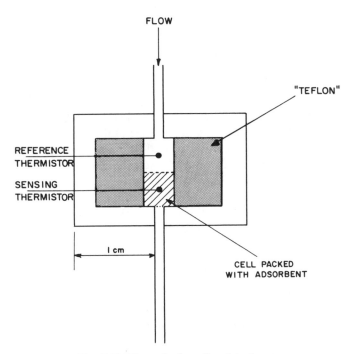

Fig. 7.12. Heat-of-adsorption detector.

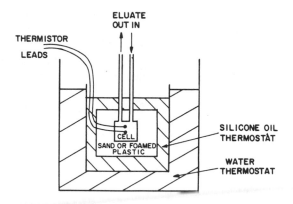

Fig. 7.13. Insulation of a heat-of-adsorption detector.

broad peaks) will have a higher response per mole and thus be as easily detected as early, sharper peaks. Detectors of this type can therefore be regarded as "universal."

A typical cell design is shown in Fig. 7.12. Eluent, containing solutes, leaves the column and flows first through the reference cell, which is empty save for the thermistor probe; immediately beyond this cell is the sensing side of the detector, which is an identical cell except that it contains the adsorbent. As with katharometer detectors in gas chromatography, so with the heat-of-adsorption system the thermistors are located in two arms of a Wheatstone bridge and, when eluent alone is flowing, variations in temperature of the cell or eluent are essentially self-compensating. Temperature changes of 10^{-4} C° are detectable; because of the variety of adsorbents and the wide range of heats of adsorption of solutes with them it is not possible to give a sensitivity level for this detector, but from literature published to date the detection of microgram amounts of solutes can be achieved.

The foregoing outline of the heat-of-adsorption detector will have raised some doubts in the reader's mind concerning, for example, the requirements for adequately thermostatting such a temperature-sensitive device and the consequence on detector output of the inevitable endothermic desorption which follows solute adsorption.

Very good environmental temperature control or, alternatively, electronic compensation for drift, is absolutely vital for the successful use of these detectors. Normal practice is to locate the cell in a heavily lagged, thermostatted–or even double thermostatted—container (Fig. 7.13).

Related to the need for good temperature control is the requirement for constant eluent flow rate, to minimize short-term instability. The various pulse-damping systems already described provide adequate smoothing even of high-pressure pulsating pump outputs.

The adsorption–desorption cycle, together with the removal of heat from the cell by the flow of mobile phase, leads to the generation of peaks of unusual form. This is illustrated in Fig. 7.14; part (a) shows the sample concentration profile as a function of time; in prat (b) we consider only the adsorption of a solute and its influence on cell temperature; if carried out adiabatically the cell temperature would rise to a higher, constant value. However, in reality the continuously flowing eluent carries the heat away from the cell and the temperature falls back to its original level, as shown. Changes associated with desorption are similar (Fig. 7.14c), but, of course, in the opposite sense. However, because the adsorption–desorption process is not instantaneous, there is a time lag between the processes, and the final detector output is shown in Fig. 7.14(d), which is the algebraic sum of the dashed temperature profiles of (b) and (c). An actual chromatogram obtained with this detector is shown in Fig. 7.15 (this figure is adapted from one published by the Japan Electron Optics Laboratory, JEOL).

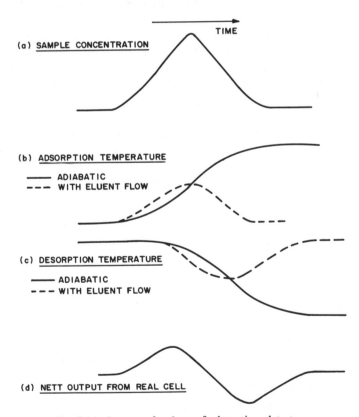

Fig. 7.14. Output of a heat-of-adsorption detector.

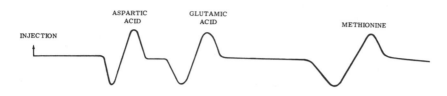

Fig. 7.15. Ion-exchange chromatogram of amino acids with heat-of-adsorption detector.

Apart from the unfamiliar appearance of this chromatogram its interpretation presents little problem. However, where poorly resolved peaks elute, the unfamiliar pattern can be hard to interpret and this has led to a reluctance to use this type of detector and attempts to modify it so that peaks of the conventional Gaussian form are obtained. Smuts *et al.*[20] sought to modify the detector so that it was essentially adiabatic; this they achieved by using a continuous feed of adsorbent into the detector cell, where it mixed with the eluent and solutes and passed over the sensor. Conditions in the vicinity of the sensor were thus virtually adiabatic, as there was no relative movement of eluent with respect to the adsorbent; conventionally shaped peaks were obtained. If the promise of Smuts's prototype adiabatic heat-of-adsorption detector can be maintained in suitably miniaturized commercial systems, the future for it looks bright.

The other disadvantage of these detectors is that gradient elution techniques are difficult because of the inevitable baseline drift associated with the adsorption of some components of the graded eluents.

In summary, heat-of-adsorption detectors are not, at present, very attractive, due to the unusual peak shape generated and thermostatting problems. Because of their universality and potential for selectivity (by careful choice of adsorbent packing for the cell), successful commercialization of an "adiabatic" variant could lead to widespread use in the future.

(c) Spectrophotometric Detectors

Spectrophotometric or colorimetric examination of fractions collected from column chromatographic separations has for long been a much used approach to their identification. Continuous monitoring of eluents, in the ultraviolet or visible regions, is now playing a correspondingly big part in the high-efficiency work we are concerned with in this book.

Considering first ultraviolet spectroscopy, there are two types of instrument available. One type is the ultraviolet spectrometer, which can measure the absorbance of a solution over a wide wavelength range from below 200 nm through into the visible region, and can scan and record complete spectra. The other type is much more limited, but considerably cheaper. This type is based on the emission from a low-pressure mercury lamp which, coupled with the use of filters and the secondary excitation of other sources by the mercury emission, provides for the measurement of absorbance at only a few specified wavelengths. Typically these are 254 nm and 280 nm, and several instruments of this type are restricted only to the former.

Instruments of both types have to be equipped with flowthrough cells and these are normally available from the instrument manufacturers.

For some purposes the cheaper, fixed-wavelength monitors are entirely adequate and they are used fairly extensively for routine analysis of proteins and nucleic acids. One can generalize to the extent that if the components of a sample are known and those that are required to be determined have adequate absorbance at the fixed wavelength(s) available, then these cheap monitors can be used. However, in research analysis, for which high-resolution column chromatography is largely used at present, such simple monitors are of limited use and the full-range spectrometers are much to be preferred. Further discussion centers on use of such instruments, though many of the comments apply equally to the simpler equipment.

The advantage of UV monitors include response, often high, to many materials of interest. It is a striking fact that many organic molecules of practical interest in biology, medicine, and industrial organic production contain aromatic groupings which, of course, absorb in the UV. Although UV detectors are not universal in response, this very selectivity to many important substances, coupled with the UV transparency (at least over significantly wide ranges) of many useful eluents makes for relatively easy use, at high sensitivity, with eluent gradients (e.g., gradients with pentane, methylene chloride, and acetonitrile can be used effectively). Ultraviolet spectra are practically independent of temperature, the sample is not destroyed, and, with full wavelength scanning, identification of eluted peaks is often possible from the spectra. In short, the UV detector has powerful advantages over, for example, the refractometer and the heat-of-adsorption detector and short column-to-detector connections with low dead-volume cells ($< 10~\mu$l) can be constructed and used effectively[21].

The principal disadvantage is the lack of universal response. Also, some desirable mobile phases are opaque to UV radiation over an important part of the wavelength range. The wavelengths at which this occurs with common eluents is given in Table 3.4.

In passing, it is worth adding that the related technique, UV fluorescence (UVF), can be used instead of simple UV absorption. The operations are identical but the fluorescence variant has two important advantages. These are its generally higher sensitivity (perhaps by one or too orders of magnitude) and the selectivity one can achieve by variation of incident radiation and full-range scanning of the fluorescence. The former advantage fits well the chromatographer's desire to use ever-smaller samples for higher resolution and also meets the need for information on ever-smaller quantities of material or on ultratrace components. The selectivity of UVF is of great value in qualitative analysis of column effluents and in differentiating between ill-resolved solutes only one of which it is desired to determine.

Photometric detectors can be used to measure absorbance in the visible region of the spectrum and for most organic solutes this requires their conversion to some colored derivative. In principle the potential of this approach is great. One has only to consider the enormous diversity of both general and selective spray reagents used in thin-layer chromatography[22] (see Table 6.2) to appreciate the possibilities. There are, however, some difficulties in the application to continuous monitoring. The two most serious ones are the often low rate of reaction, which would at least require that the eluate and reagent be heated, and the serious band broadening that could arise through introduction of reagent and the diffusive mixing which would result. These the problems are not insuperable and growing use of this approach, particularly in routine analysis, can confidently be predicted. At present, though, the only widely used commercial continuous monitor available of this type is the amino acid analyzer based on reaction with ninhydrin.

To summarize, photometric detectors are selective and frequently have high response. They are little affected by environmental changes and so are rather easier to use than many other detectors, though they tend to be expensive. By use of appropriate, and rapid, chemical conversions photometric detectors may, in the future, be made selectively sensitive to almost any type of organic solute.

(d) Transport-Ionization Detectors

It is first necessary to explain the cryptic name given to this class of detector. The essential components of a system of this type are shown sche-

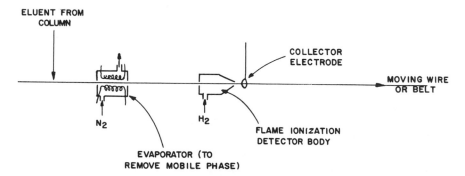

Fig. 7.16. Elements of a transport-ionization detector.

matically in Fig. 7.16. Eluate from the column is picked up on a moving metal wire, chain, or belt. It passes through a heated chamber, usually purged with nitrogen, to strip off the large excess of volatile mobile phase. The less volatile solute remains on the metal "transporter" and is carried to an ionization detector of a type commonly used in gas chromatography.[23] In some designs the transporter passes directly through a hydrogen flame; in others the solute is pyrolyzed and the pyrolysis products flushed into an ionization detector (i.e., argon, flame, or electron capture). Thus the essentials are means of transport of successively eluted peaks, stripping of mobile phase, and high-sensitivity ionization detection of the solutes.

At the time of writing at least five commercial instruments of this type are available, one of which offers a choice between the different ionization detectors. Before discussing the variations among the different designs it is worth considering the significant advantages this type of detector has over others which have so far been described. These advantages are such that, with careful selection of the best of the possible design variations, this type of detector might well become the much sought after "universal" instrument, at least for hydrophobic samples of boiling point above about 250°C.

The three advantages which particularly distinguish this system are (i) its lack of response to the mobile phase (which should never get to the detector), (ii) its insensitivity to temperature, flow rate, and other environmental factors, and (iii) its very high sensitivity to solutes. These detectors are fairly expensive, reasonably easy to use, destructive of a proportion of the solute, and make a relatively small contribution to solute band spreading. Virtually all of the above comments on these detectors are half-truths in the sense that of the different units on the market, some are more successful than others in achieving the full potential of the advantages enumerated. At present there must be a little doubt about the quantitivity of these detectors. This arises because the transporter carries only a small fraction of the eluate to the detector. The assumption that this fraction is constant and also independent of solute (and mobile phase composition in gradient elution) has not yet been positively demonstrated.

The features we will consider in more detail are: the means of transferring eluate to the "transporter," the nature of the "transporter" system, the solvent evaporator, the means of transferring solute from the "transporter" to the ionization detector, the ionization detector, and the cleaning of the transporter.

Application of eluate to the transporter is achieved by contacting the stream, or drops, of liquid issuing from the end of the column with the transporter. The latter, moving continuously past the column exit, carries a more or less constant proportion of the eluate to the detector. The proportion of the eluate collected on the transporter depends on the form and geometry of the latter, which therefore has an influence on the sensitivity. The amount collected is also affected by the viscosity and surface tension of the eluate and by the "wettability" of the metal. Actual contacting of the liquid may be achieved by drawing the transporter through the drop of liquid at the exit of the column, but particularly with the chain and belt types of transporter (see next paragraph), the eluate is frequently allowed to drip onto the moving metal. That part of the eluate not carried away by the transporter can be recovered in a fraction collector for further examination.

Important features of the transporter system include the material of construction, the physical form (e.g., wire, chain, or belt), and the means of provision of continuous transport (e.g., a looped and continuously circulating or a very long, one-pass system). Generally, looped systems are of the chain or belt type and are fabricated from platinum or gold. The long, one-pass system is normally based on an iron wire. Between the two systems there are some significant practical differences. A major disadvantage of the "endless" chain or belt is that any material difficult to burn or pyrolyze which becomes deposited onto the transporter contributes a signal at each pass through the detector. Chain transporters are particularly prone to this "noise" problem, as material tends to become trapped between the links and is only slowly lost. In addition fracture of chains has been found to occur sufficiently frequently to be rated an objection. Important advantages of the "endless" transporter are that (i) because it is short, it can be constructed of inert but costly materials such as platinum or gold, and (ii) that it is able to carry a proportion of the total eluate several times greater than a single wire can accommodate. In the extreme one could envisage even total transfer of eluate, using a belt with a trough form of cross section. The main limiting factor on the quantity of eluate transported, apart from the actual capacity of the transporter, is the necessity of preventing band broadening due to diffusion, or even more seriously, swirling or flowing of the eluate on its transit from the column exit to the solvent evaporator. Nevertheless it seems possible that the limited volume of eluate from microcolumns could be transported quantitatively to detectors of this type.

Among the features of the wire type of transporter not yet discussed are its advantageous low cost (for iron wire of suitable gauge the current price is about $10 for 20,000 m). The running time of a spool of wire (typical wire speed $\sim$6 cm sec^{-1}) is 90 hr so that many analyses, or alternatively, very long separations, can be monitored with a single uncontaminated transporter. As usual this advantage is not realizable to the full in practice because in their production, wires are "drawn" through oil-lubricated dies and small amounts of the organic matter become occluded in the metal. This is partly realized on passage of the wire through the pyrolysis furnace, giving rise to detector "noise." Some workers have used direct resistive heating of the wire (to red-heat) well upstream of the eluent pick-up point to burn out the contaminating oil, while others have run unused wires through the system to achieve significant clean-up prior to actual use.

The problem of transporter wettability has been mentioned earlier. Probably because their capacity, related to surface area, tends to be small, the shortcomings of iron transporters are sometimes apparent in this respect, although in the writers' experiences the problem is much less obvious with organic than with aqueous mobile phases. One company marketing a wire-transport detector specifically states that, "the surface of the sampling wire has been modified so that it is wetted by aqueous and organic solvents. The resulting sampling process is smooth and continuous." The most likely, inexpensive way in which this could be achieved would be to roast the wire in air to form a somewhat porous, adsorptive oxide "skin."

Although in principle the solvent evaporator is a simple device—a heated, gas-purged chamber—the reader will not be surprised to learn that there are practical problems in its use. The critical decision to be made is the temperature at which to operate the evaporator. The residence time (and therefore in most cases the length of the evaporator) is a secondary factor. The objective is to remove mobile phase completely while retaining all of the solute on the transporter. The simplest case is where a really volatile eluent, pentane perhaps, has been used to separate polymers. Here the volatility difference is enormous and one might think the stated objective could be achieved with ease. However, this is not always so: too rapid vaporization leads to sputtering and consequent loss of solute. Rather like the too rapid drying of paint, involatile solute skins on the transporter may blister and flake away. In addition polymers tend to retain traces of solvents tenaciously. Consequently, the use of these detectors with polymeric solutes does pose problems. At the other extreme the more volatile solutes (e.g., those boiling below about 300°C) will at least partially be lost while the eluent is being evaporated so that, in effect, sensitivity will become low. Between these extremes there is a wide range of solutes which can be handled very adequately by transporter-type detection systems.

As we proceed through the detection system we have now reached the point where the solute (eluent-free) is on the transporter. There are two systems commercially available for transferring it to the actual ionization detector: a third, potentially superior approach has been described and may well soon become available. The first and most direct system is to cause the laden transporter to pass directly through the flame of a flame ionization detector. The solute is burned, probably completely, and the ions generated collected in the usual way. This system has been criticized on account of the flame noise consequent upon the direct movement through the flame. On top of this inherent noise involatile, carbonaceous contaminants also burn; where an endless-belt/chain transporter is used and is thus contaminated, a regular signal appears superimposed on the "true" chromatogram. These criticisms are supported by observations in the authors' laboratory.

The second, commercialized solute transfer process is based on pyrolysis from the wire, belt, or chain. Most organic material is degraded when heated to temperatures above 500°C and the fragments formed are readily volatile (for most organics, pyrolyzed for a few seconds at $\sim$700°C, probably over 90% of the pyrolysis products boil below 150°C). Thus, by pyrolyzing solutes from the transporter into a stream of gas which carries the fragments into an ionization detector the noise problem of the direct flame-transit transporter system is avoided.

The third system referred to is a variant on the previous one. combustion, not pyrolysis, is used. All the organic matter on the transporter is burned off; the combustion products are then passed over heated carbon and, after admixture with hydrogen, over hot nickel, prior to sweeping into the flame (or argon) ionisation detector. The reactions occurring are

$$C_xH_y + \ldots \xrightarrow{\ O_2\ } xCO_2 + \ldots$$

$$xCO_2 \xrightarrow[\text{carbon}]{\text{hot}} 2xCO$$

$$2xCO \xrightarrow[\text{hot nickel}]{H_2} 2xCH_4$$

and the consequence is that for each mole of carbon dioxide originally formed, two of methane are produced; methane is of course detectable at very low levels with ionization detectors. Martin and Scott[24,25] have described this system, called a "molecular multiplier," and have indicated its potential for increasing sensitivity of detection through a sequence of oxidation and reduction steps. Each cycle doubles the amount of detectable material, so that three orders of magnitude of sensitivity increase can theoretically be obtained from ten such cycles. Application of such processes can be expected in the future to lead to ultrahigh-sensitivity detectors responding quantitatively to the amount of carbon present in a detected solute.

The last two features of the "transport-ionization" detectors to be discussed can both be dismissed quickly.

First, choice of the ionization detector. Although the argon detector was the first to be used in systems for liquid chromatography, it is being superseded by the flame ionization device. The relative merits of these detectors are well-documented in textbooks on gas chromatography and the reader is referred to such books for further information. Essentially, either detector gives a universal response to organic matter; one can anticipate the introduction of reactors (like the carbon–nickel systems of the molecular multiplier) to convert certain components to a form giving a high response in one or other of the selective detectors coming into prominence in gas chromatography. For example, pyrolytic hydrogenolysis of solute from the transporter into a flame photometric or microcoulometric detector might be used to produce "sulfur chromatograms" based on the detection of hydrogen sulfide.

Particularly with the endless-belt transporters, it is necessary to clean any remaining material from the belt or chain after it has carried solute to the detector. To date only relatively unsophisticated heating and solvent washing steps have been used and their effectiveness is variable.

We have devoted a considerable amount of space to these transport-ionization detectors, highlighting their versatility and their potential for further refinement and extension of applicability. This is because we believe such detectors offer the best combination of sensitivity, universality of response, and ease of use of all the types so far described and applied to the monitoring of high-efficiency liquid chromatographic separations.

(e) Radioactivity Detector

Isotopic labeling techniques have been used widely in chemical and biochemical research for at least two decades and a detector to monitor labeled compounds separated by gas chromatography was described[26] and marketed some years ago. More recently a liquid scintillation counting system has been developed for use as a liquid chromatography detector for ^{14}C- and ^{3}H-labeled compounds (see the appendix). The chromatographic effluent passes through a glass tube filled with a scintillator, such as powdered anthracene. Suitably placed photomultiplier tubes detect the flashes of light produced as β-particles impinge onto the scintillator and the resultant photomultiplier output is displayed to provide the chromatogram.

Detectors of this type can be regarded as very specialized, only applicable to a rather sophisticated research objective and not, therefore, likely to be generally useful. However, the very high sensitivity of radio-activity detectors in general makes them worthy of further consideration.

Disintegration rates of the order 100 min^{-1} can readily be detected. This is equivalent to a detection limit for ^{14}C or ^{3}H of around 10^{-12}–10^{-14} g of the radionucleide, a very high sensitivity indeed compared with most existing detectors. Taking this fact in conjuction with the rather facile hydrogen exchange which occurs with many substances, the possibility of using radio-activity detectors more widely begins to look promising.

The exchange of protons in hydroxylic or carboxylic groups is well-known; protons in aromatic nuclei will exchange with those of tritiated sulfuric acid:

$$\text{\textcircled{$-$}}H + {}^{3}HHSO_4 \rightleftharpoons \text{\textcircled{$-$}}{}^{3}H + H_2SO_4$$

Thus it seems likely that many complex samples may be tritiated by rather simple chemical exchange reactions prior to chromatography and may then be detected after separation by very high-sensitivity radioactivity detectors. It could be that, if the sensitivity of more conventional detectors cannot be further improved, attention may be directed to the use of radioactivity as the most sensitive monitor of high-efficiency LC of submicro samples.

Apart from its high sensitivity to labeled substances the advantages of a radioactivity detector are its complete insensitivity to the mobile phase (unless exchange between it and the labeled solute occurs) and to ambient changes. On the other hand, in addition to obvious disadvantages of limited applicability, devices of this type are expensive.

(f) Electrical Conductivity Detectors

Conductance measurements have for long played an important role in the determination of substances, particularly in aqueous solutions. A typical conductance probe consists of two plane-parallel platinum electrodes immersed in the solution to be measured and electrically forming an arm of a Wheatstone bridge. By replacing the normal probe by a low-volume flow-through cell with small, carefully mounted electrodes, and with recent improvements in measuring circuit technology, conductivity monitors suitable for use with aqueous and nonaqueous column chromatography are becoming available.

Descriptions of several[27-29] conductivity detectors have been published, some of which are either of particularly low volume or high sensitivity. Naturally they are all primarily designed for, and evaluated with, aqueous systems. The possibility exists that these or improved versions may be applicable to nonaqueous LC (with which we are principally concerned in this book). In a most interesting article[30] on conduction of polar liquids, Bright and Makin note the very considerable increase which can be achieved in the resistivity of polar liquids by modern purification methods, including the use of ion-exchange resins. For example, the resistivity of nitrobenzene increases from about 2×10^6 ohm m^{-1} to about 10^8 ohm m^{-1} by treatment with these resins for a few minutes. It is possible that other materials used as chromatographic stationary phases may also bring about similar increases in the resistivity of organic liquids. One draws two conclusions from these observations. First, mobile phases and the bulk of most organic samples used in liquid chromatography are very poor conductors indeed. Second, the trace polar components shown to be removable from organic materials probably have much higher conductivity. Thus it seems possible that electrial conductivity could become a very useful method of monitoring the separation of trace components of organic materials. This would be of considerable value since such components may play an important part in the generation of dangerous electric charge build-up in flowing organic streams, such as petroleum products in pipelines.

The above is a wholly speculative view of the extension of use of a group of detectors which are now available and widely and successfully being used for the monitoring of many inorganic and biochemical separations by column LC.

Their chief advantages, in present usage with largely aqueous eluates, are lack of response to mobile phase, fairly high response which is predictable from conductance data, insensitivity to modest changes in temperature and flow rate, nondestructive nature, and, of considerable importance, ease of construction and operation. Although most electrical conductivity detectors are usually claimed to have a linear response, the low-volume cell described by Pecsok and Saunders[27] did not, and it is possible that in any extension of application to partly or wholly nonaqueous media nonlinearity may be a serious limitation.

7.9.3. Experimental Detection Systems

In addition to those systems for continuous effluent monitoring which have been commercialized, there are a number of others which have been described which seem to have real value. It is likely some of these may become available commercially before this book is published.

(a) Polarographic Detectors

Compounds which undergo electrochemical reactions can be detected by the polarographic technique and Kemula has for some years been using a polarographic detector to monitor liquid chromatographic separations. The system, Kemula called it "chromatopolarography," was first described in 1952[31] and subsequent papers from the same author have demonstrated its applicability to nitro compounds and DDT isomers[32], amino acids[33], alkaloids[34], and aldehydes and ketones[35]. Kemula has given a recent account of his work in an easily accessible Western journal[36]. The work referred to did not involve the fast or high-performance liquid chromatography which is coming into prominence now, but did serve to demonstrate that small quantities of an interestingly wide range of organic molecules could be detected.

Koen et al.[37] have described the construction, operating characteristics, and application of a low-cost polarographic detector specifically designed for use with high-efficiency columns. Key features of this system are low dead volumes between column exit and detector exit (including, of course, the polarographic cell), a more rapid drop time than is normally used in polarography (in order to measure fast eluted peaks more precisely), and an electronic circuit providing more selective damping than normal to combat the current fluctuations which arise from the high frequency of the mercury drops. The detector itself is of simple construction (see Fig. 7.17). The sample concentration detection limit for this detector was found to be about 10^{-8} mole liter^{-1} for p-nitrophenol, methyl parathion, and parathion itself.

The polarographic detector does not respond universally to solutes and is sensitive to electroactive impurities, such as oxygen or metal ions. Further, it requires an eluent of reasonably high conductivity, so that the choice of mobile phase for organic analysis is restricted. Thus its applicability in organic chemistry is somewhat limited, but its relatively high and selective response to some solutes, coupled with the low dead volume, suggest that it may prove to be a valuable selective detector in the future.

(b) Heat-of-Reaction Detector

This system of detection has points in common both with the heat-of-adsorption detector and the photometric detector. The essence of the system is to add a reagent to the column effluent and measure the temperature

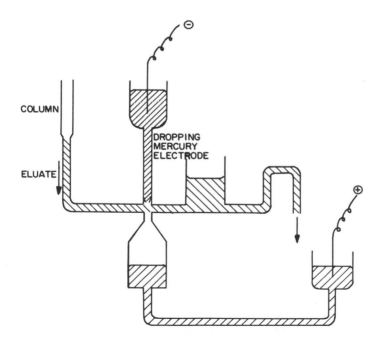

Fig. 7.17. A polarographic detector.

change accompanying any reaction. A system of this kind for continuous monitoring of a plant stream (rather than a chromatographic effluent) has been described[38]. In the application, triethylaluminum reacted with various impurities (water, dissolved oxygen, and alcohols) in a hydrocarbon stream. Reaction takes place in a carefully thermostatted cell ($\pm 0.002°C$) and the heat generated is measured by a thermistor.

Application of this system provides selective detection, and, as with the photometric detectors, range of selectivity is wide, the major problem being to find a reaction rapid enough to give essentially complete reaction in times up to about 30 sec. Compared with the heat-of-adsorption detector one notes the important feature they have in common, the need for very good thermostatting or, alternatively, electrical compensation for temperature changes. The differences between the two detectors are important, too. Reactions are chemically selective, whereas adsorption is a much more general phenomenon. Thus the adsorption detector is "universal," though at the same time sensitive to changes in the mobile phase; the reaction detector's selectivity should make it very useful for routine analysis once a suitable column separation has been developed. Another significant difference between the two detectors is that the reaction system gives conventional, Gaussian peaks as distinct from the "differential" peaks associated with adsorption–desorption behavior in the other detector. Heats of reaction are normally considerably greater than heats of adsorption, so that reaction detectors are potentially the more sensitive. The main exception to that generalization concerns chemisorption, as, for instance, in certain adsorptions on ion-exchange resins, where the heat of sorption is of a similar magnitude to heats of reaction. Finally, because the reaction detector requires the addition and mixing of a reagent, the cell design is critical and may contribute significantly to band broadening; in this respect reaction and colorimetric detectors are both at a disadvantage compared with heat-of-adsorption systems.

To summarize, the authors believe that heat-of-reaction detectors (like colorimetric ones) are capable of development to provide sensitive, selective detection with a wide variation in selectivity provided by one reaction cell and an armory of suitable reagents which can be fed to it.

(c) Density Detectors

A prototype densitimetric detector which records the very slight changes in effluent density that occur as solutes are eluted from an LC column has been exhibited[39]. The column effluent flows through a cell containing a polytetrafluoroethylene "diver" in which a magent is embedded. An externally mounted magnet holds the diver at a suitable height above the cell bottom and in such a position that it interrupts a beam of light falling through the cell onto a photocell. As the effluent density changes, the diver position varies slightly and the amount of light falling on the photocell changes accordingly. Sensitivity of this detector was claimed to be about 0.002 g ml^{-1} change in effluent density. Common mobile phases have densities in the range 0.65–0.9 g ml^{-1} and solutes like alkaloids, carbohydrates, nitro compounds, and the larger organic molecules in general have densities between 1 and 2 g ml^{-1}. The claimed sensitivity of the prototype would, for solute of density 1.2 g ml^{-1} and mobile phase density 0.9 g ml^{-1}, enable detection of a minimum of about 1000 ppm of solute. This is two to three orders of magnitude less than that needed for the highest-resolution work and it seems unlikely that density detectors will be capable of improvement to the requisite degree. Since the diver-in-cell system does not lend itself too easily to minimizing peak broadening, it is concluded that it is unlikely to be widely used in the future.

(d) Dielectric Constant Detectors

By maintaining a high potential between close-mounted parallel or concentric elements, changes in the dielectric constant (i.e., electrical polarizability) of a liquid flowing between the elements may be measured. This is the basis of dielectric constant detectors. In almost every way dielectric constant is the electrical analog of refractive index; there is, however, one feature which, chromatographically, makes the measurement of this property less attractive than measurement of refractive index. That feature is the close parallel which exists between dielectric constant and eluotropic series ("polarity"), a situation which does not obtain to any significant extent with refractive index. Reference to Table 7.4 reveals that n_D for the series pentane, toluene, diethyl ether, acetone, isopropanol varies only slightly, except for toluene, where the presence of an aromatic ring leads to a much higher refractive index than for the other substances. The dielectric constants, however, increase progressively (pentane = 1.844, toluene = 2.379, diethyl ether = 4.335, isopropanol = 18.3). Since in liquid adsorption and, to some extent, partition chromatography the ideal eluent has a polarity similar to that of the solutes to be separated (hence, for example, use of gradient elution), dielectric constant differences tend inevitably to be minimized by chromatographic requirements. The loss in potential sensitivity on this score may be only partially nullified by the fundamentally greater precision of measurement of electrical, compared with optical, properties of solutions. Too little is known about the design and performance of dielectric constant detectors to predict their future applicability with any confidence, but they do not seem to offer any significant advantages over detectors already commercially available.

(e) Vapor Pressure Detectors

Several groups of workers have described continuous vapor pressure detectors[40-43] which are based on the same principle as the widely used vapor pressure osmometers, for measuring number average molecular weights of involatile solutes. Poulsen and Jensen[43], in particular, explored the performance of such a detector for liquid elution chromatography, seeking to optimize sensitivity and minimize detector dead volume. Their detector had a limit of detection of the order 10^{-5} g solute of molecular weight 200 in a 1-ml bandwidth and a minimum dead volume of about 10 μl; the dead volume did, however, increase with increasing flow rate. The sensitivity is therefore approaching that required for low-sample-loading, high-resolution work and the dead volume is remarkably low under favorable flow circumstances. The advantages of a detection system of this type include a virtual universality of response, which is predictable and linear, and the fact that it is that nondestructive. Disadvantages are that the solutes need to be substantially less volatile than the solvent and the rather large dead volume at high flow rates (e.g., 90 μl at 2 ml min^{-1}), which might well preclude its use for fast analysis, where one is trading column performance for speed and using flow rates of the order 1 ml min^{-1}.

APPENDIX: A SHORT GUIDE TO COMMERCIAL EQUIPMENT SUITABLE FOR COLUMN LIQUID CHROMATOGRAPHY

Sieves

A. Fritsch OHG, 6580 Idar-Oberstien 1, Haupst-rasse 542, W. Germany	Conventional sieves, with water washing facilities
Endecotts Ltd., Lombard Rd., London S.W.19, G.B.	Conventional sieves down to 5.5 μm
Veco N. V., Karel van Glereweg 22, Eerbeek, Netherlands	Electroformed sieves down to 5 μm
J. Engelsmann A.G., Frankenthaler Strasse 137–141, 67 Ludwigshaven, W. Germany	Conventional sieves down to 25 μm

Grinding Equipment

A. Fritsch OHG, 6580 Idar-Oberstein 1, Haupst-rasse 542, W. Germany	Powered pestle and mortar
G.E.C. (Engineering) Ltd., Fraser & Chalmers Eng. Works, Erith, Kent, G.B.	High-frequency vibratory mill

Pumps

	Type	Maximum pressure, atm
Cenco Instruments N.V., Konijnenberg 40, Breda, Netherlands.	Plunger	66

Phoenix Precision Instruments Co., Inc., 3803 N. Fifth St., Philadelphia, pa.	Gradient pump	66
Metering Pumps Ltd., 49–51 Uxbridge Road, Ealing, London W.5., G.B.	Plunger	660
Bran and Luebbe (G.B.) Ltd., 62 Coventry Rd., Market Harborough, Leicester, G.B.	Plunger	2000
Orlita K.G., 63 Giessen-Lahn, Max Eyth Strasse 10, W. Germany	Diaphragm	325
Waters Associates Inc., 61 Fountain Street, Framingham, Mass. 01701	Pulseless	~10
Milton Roy Co., St. Petersburg, Fla. 33733	Various types	500
Whitey Research Tool Co., 5679 Landregan St., Emeryville, Calif. 94608	Diaphragm	350
F. A. Hughes & Co. Ltd., Dienhcim Road, Epsom, Surrey, G.B.	Plunger	150
Technicon Corp., Chauncey, New York,	Plunger	66
Nester/Faust Inc., 2401 Ogletown Road, Newark, Del. 19711	Pulseless piston	100
Isco, 4700 Superior, Lincoln, Neb. 68504	Diaphragm	5

Pulse-Damping Equipment

Gauges-Bourdon (G.B.) Ltd., Bramley, Guildford, Surrey, G.B.	Bourdon tube
Fawcett Preston & Co. Ltd., Bromborough, Cheshire, G.B.	Greer–Mercier hydropneumatic accumulators

Sampling Valves

	Capacity, μl	Maximum pressure, atm	Maximum temperature, C
Seiscor, Box 1590,	0.5	100	150
Tulsa, Okla.		25	220
Valco Insts., P.O. Box 19032, Houston, Tex. 77024	0.2–50	140	300
Gervase Insts. Ltd., 3 Kingsway Gdns., Chandlers Ford, Hants., G.B.	1–50	14	200
Tracor Inc., Analytical Instrument Division, 6500 Tracor Lane, Austin, Tex. 78721	0.25–10	66	220

Detectors

	Types available*
LKB-Produkter A.B., P.O. Box 76, Stockholm-Bromma, Sweden	c,g
Instrumentation Specialties Co., Inc., 4700 Superior, Lincoln, Neb. 68504	c
Nester/Faust Inc., 2401 Ogletown Rd., Newark, Del. 19711	a,b,c,e,g
E.I. du Pont de Nemours, Wilmington, Del. 19898	a,d

*Code for detector types: a. refractive index; b. heat of adsorption; c. ultraviolet or visible (selected wavelengths only); d. ultraviolet or visible; e. transport-ionization; e. f. radio-activity; g. electrical conductivity.

Carlo Erba, Via Carlo Imbonati, 24, Milan, Italy	e
Varian-Aerograph Inc., 2700 Mitchell Drive, Walnut Creek, Calif. 94598	a,b,c
Waters Associates Inc., 61 Fountain St., Framingham, Mass. 01701	a,c
Cenco Instruments Corp., Konijnenberg 40, Breda, The Netherlands	c
Technical Measurement Corp., 441 Washington Ave., North Haven, Conn. 06473	d
Japan Electron Optics, Laboratory, Co. Ltd., 1418 Nakagami, Akishima, Tokyo, Japan	e
E-C Apparatus Corp., University City, Philadelphia, Pa. 19104	a
N.V. Philips Gloeilampenfabrieken, Eindhoven, The Netherlands	f
Barber-Colman Inc., Rockford, Ill. 61101	e
Pye-Unicam Ltd., York St., Cambridge, G.B.	e
Packard Instrument Co., Inc., 2200 Warrenville Rd., Downers Grove, Ill. 60515	e
Chromatronix, Inc., 2743 Eighth St., Berkeley ,Calif. 94710	g
Laboratory Data Control, 42 Shelter Road, P.O. Box 1030 Danbury, Conn. 06810	a,c

Complete Liquid Chromatographs

Waters Associates, Inc., 61 Fountain St. Framingham, Mass.
Du Pont Instruments, Inc., Wilmington, Del. 19898
Nester/Faust Manufacturing Corp., 2401 Ogletown Road, Newark, Del. 19711
Varian Aerograph, 2700 Mitchell Drive, Walnut Creek, Calif. 94598
Siemens Aktiengesellschaft, D7500 Karlsruhe 21, P.O. Box 211080, W. Germany
Problematics Inc., 223 Crescent St., Waltham, Mass. 02154

REFERENCES

1. 1969–1970 Laboratory guide, *Anal. Chem.* July 1969.
2. International chromatography guide, *J. Chromatog. Sci.,* May 1969.
3. Johnson, J. F., and Cantow, M. R., *J. Chromatog.* **28,** 128 (1967).
4. Snyder, L. R., and Saunders, D. L., *J. Chromatog. Sci.* **7,** 195 (1969).
5. Rosett, T., *J. Chromatog.* **18,** 498 (1965).
6. Middleton, W. R., *Anal. Chem.* **39,** 1839 (1967).
7. Hurrell, R. A., and Amos, R., in *Gas Chromatography 1962,* van Swaay, M., ed., Butterworths, London p. 162, (1963).
8. Scott, R. P. W., Blackburn, D. W. J., and Wilkins, T., *J. Gas Chromatog.* **5,** 182 (1967).
9. Stewart, H. N. M. S., Amos, R., and Perry, S. G., *J. Chromatog.* **38,** 209 (1968).
10. Huber, J. F. K., personal communication.
11. Halasz, I., and Walking, P., *J. Chromatog. Sci.* **7,** 129 (1969).
12. Maggs, R. J., and Young, T. E., in *Gas Chromatography 1968,* Harbourn, C. L. A., ed., Institute of Petroleum, London p. 217, (1969).
13. Snyder, L. R., *J. Chromatog. Sci.* **7,** 352 (1969).
14. Giddings, J. C., *Dynamics of Chromatography,* Part I, Marcel Dekker, New York, p. 281, (1965).
15. Piel, E. V., *Anal. Chem.* **38,** 670 (1966).
16. Snyder, L. R., *Anal. Chem.* **39,** 698 (1967).
17. Huber, J. F. K., *J. Chromatog. Sci.* **7,** 172 (1969).
18. Billmeyer, F. W., Jr., and Kelley, R. N., *J. Chromatog.* **34,** 322 (1968).
19. Bombaugh, K. J., King, R. N., and Cohen, A. J., *J. Chromatog.* **43,** 332 (1969).
20. Smuts, T. W., van Niekerk, F. A., and Pretorius, V., *J. Chromatog. Sci.,* **7,** 127 (1969).

21. Kirkland, J. J., *Anal. Chem.* **40**, 391 (1968).
22. Stahl, E. (ed.), *Thin Layer Chromatography*, 2nd English ed., Springer-Verlag, Berlin, (1969).
23. Littlewood, A. B., *Gas Chromatography*, 2nd ed., Academic Press, London and New York (1970).
24. Martin, A. J. P., and Scott, R. P. W., Las Vegas Symposium, 1969.
25. Scott, R. P. W., in *Gas Chromatography in Biology and Medicine*, Porter, R., ed., Churchill, London, pp. 54–56, (1969).
26. James, A. T., and Piper, E. A., *J. Chromatog.* **5**, 265 (1961).
27. Pecsok, R. L., and Saunders, D. L., *Anal. Chem.* **40**, 1756 (1968).
28. Knudson, G., Ramaley, L., and Holcombe, W. A., *Chem. Instr.* **1**, 325 (1969).
29. Taylor, A. F., *Analyst* **88**, 145 (1963).
30. Bright, A. W., and Makin, B., *J. Mat. Sci.,* **2**, 184 (1967).
31. Kemula, A., *Roczniki Chem.* **26**, 281 (1952).
32. Kemula, A., *J. Anal. Chem. (USSR)* **22**, 562 (1967).
33. Kemula, W., and Witwicki, J., *Roczhiki Chem.* **29**, 1153 (1955).
34. Kemula, W., and Stachurski, Z., *Roczhiki Chem.* **30**, 1285 (1956).
35. Kemula, W., Butkiewicz, D., and Sybitska, D., *Modern Aspects of Polarography*, Plenum Press, New York p. 36, (1966).
36. Kemula, W., and Sybitska, D., *Anal. Chim. Acta* **38**, 97 (1967).
37. Koen, J. G., Huber, J. F. K., Poppe, H., and den Boef, G., *J. Chromatog. Sci.* **8**, 192 (1970).
38. Crompton, T. R., and Cope, B., *Anal. Chem.* **40**, 274 (1968).
39. Physics Exhbition, London 1967, Chemical Inspectorate of the Ministry of Defence exhibit.
40. Sternberg, J. C., and Carson, L. M., *J. Chromatog.* **2**, 53 (1959).
41. Simon, W., Clerc, J. T., and Dohner, R. E., *Helv. Chim. Acta* **48**, 1628 (1965).
42. Simon, W., Clerc, J. T., and Dohner, R. E., *Microchim. J.* **10**, 495 (1966).
43. Poulsen, R. E., and Jensen, H. B., *Anal. Chem.,* **40**, 1206 (1968).

FURTHER READING

Conlon, R. D., Liquid chromatography detectors, *Anal. Chem.* **41** (4), 107A (1969).

Halasz, I., and Walkling, P., Different types of packed columns in liquid–solid chromatography, *J. Chromatog. Sci.* **7**, 129 (1969).

Horvath, C., and Lipsky, S. R., Rapid analysis of ribonucleosides....at the picomole level using pellicular cation exchange resin in narrowbore columns, *Anal. Chem.* **41**, 1227 (1969).

Horvath, C., and Lipsky, S. R., Column design in high-pressure liquid chromatography, *J. Chromatog. Sci.* **7**, 109 (1969).

Huber, J. F. K., High-efficiency, high-speed liquid chromatography in columns, *J. Chromatog. Sci.* **7**, 85 (1969).

Hupe, K. P., and Bayer, E., A microadsorption detector for general use in liquid chromatography, *J. Gas Chromatog.* **5**, 197 (1967).

Jentoft, R. E., and Gouw, T. H., A high-resolution liquid chromatography, *Anal. Chem.* **40**, 1787 (1969).

Maggs, R. J., Application of a F.I.D. as a monitor for liquid chromatography columns, *Column (Pye Unicam)* **2** (2,d) (1967).

Veening, H., *J. Chem. Ed.* **47**, (1970). The issues for September, October, and November contain a review of detectors for LC.

The papers and reviews listed above typify the recent work in the field of high-speed, high-resolution liquid chromatography which has led to to the revolution in liquid chromatographic technique.

Chapter 8

Current Status of Liquid Chromatography: Place Among Family of Chromatographic Techniques

Chromatography has reached a particularly interesting stage at the start of the 1970's. Whereas a decade ago gas chromatography could fairly be claimed to stand supreme in terms of performance and ease of use, TLC, and soon afterwards gel permeation chromatography, were beginning to have an impact. More recently, developments which have led to major improvements in column LC and the introduction of chromatography with supercritical mobile phases have further widened the range of separation tools available to the analyst. In this chapter we want to look at these various tools and see where their applicability overlaps and what criteria should be applied in choosing between the alternatives. We will also note the likely developments which may influence the choice in the future.

Gas chromatography is unquestionably the most sophisticated chromatographic technique presently in use, but nevertheless it is extremely simple to apply because the instrumentation required is well-developed. Consequently, attempts are made to use it for the separation and analysis of a very wide range of materials. In many cases such attempts are entirely successful, no unforeseen complications arise, and accurate, precise analyses are obtained. However, there are some applications where one has rather less confidence in the value of the results obtained. One thinks particularly of the analysis of relatively involatile substances, for example, pesticides and steroids, and of materials which are reactive or chemically or thermally labile. In these cases sometimes efforts are made to convert the materials to a form more suitable for effective, accurate gas chromatography. Pyrolysis is widely used in order that volatile fragments from involatile substances (such as polymers) may be chromatographed and thus provide analytical information otherwise unobtainable. Esterification and silylation are used to convert polar compounds, which might otherwise react or become adsorbed on the column, into less polar derivatives. These derivatives are often more volatile than their precursors, so that the conversion is beneficial in two ways–as, for example, the ability to chromatograph silyl derivatives of carbohydrates.

The degradation of samples in gas chromatographs is well-documented. Reactivity of solutes toward the column wall material, the solid support, and precolumn or postcolumn piping at elevated temperatures are usually the cause. Samples may be fragmented (as, for example, saturated hydrocarbons to olefins in GSC on molecular sieves at $> 350°C$), partially, or wholly adsorbed (e.g., some sulfur compounds on copper or other metal tubing), or isomerized (e.g., terpene conversions).

Until quite recently the technology and instrumentation of GC have been so far in advance of any potentially competitive technique that these hazards have been accepted and, in many cases, successfully overcome. Nevertheless, the advent of new alternatives to GC make reexamination of some present GC applications worthwhile. For example, derivatization methods prior to GC can be time-consuming; methods enabling avoidance are attractive.

Liquid chromatography holds promise of surplanting GC for some important separations of less volatile materials. The Ciba Foundation held a symposium in February 1969 on gas chromatography in biology and medicine[1]. The discussions there, among the foremost chromatographers of the day, make stimulating reading, and the following are particularly relevant:

Lipsky: "I predict that the trend in chromatography will soon turn toward LC again. Systems in which picomole quantities can be analyzed will become commonplace. We shall eventually be able to analyze steroids, including the sex hormones, with these sorts of techniques."

Scott: "I agree ... we are in a transitional stage between GC and LC and many wokers are still trying to use GC techniques beyond the practical limits of their operation. Until 2–3 years ago there was no alternative to GC, but recent work shows that this is no longer the case. The extent to which we shall continue to use GC at high temperatures and tolerate all the associated problems is limited."

The discussion continued in this vein and received general, but not quite unanimous support. Let us look at some of the work which is being done with LC today.

Figure 8.1 is a liquid chromatogram of ribonucleosides mono-, di-, and tri-phosphoric acids obtained by Horvath et al.[2] using a 2 m × 1 mm

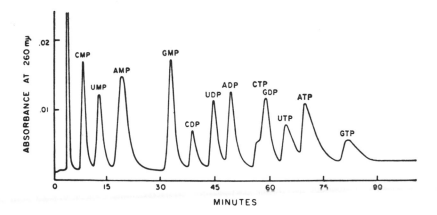

Fig. 8.1. Separation of ribonucleosides mono-, di-, and tri-phosphoric acids (by courtesy of *Analytical Chemistry*).

strong anion-exchanger column with small-diameter ($\sim$50 μm) resin particles operated at high pressures (up to 300 atm). There are several features of interest in this work. The analysis time of about 90 min compares with about 20 hr for the commonly used column chromatographic method, which also requires appreciably larger samples. The analysis time is of the same order as for a complex GC analysis, the columns are reusable, and a continuous record of the separation is obtained. Nucleic acid hydrolysates, like this sample, are thermally unstable but can be readily handled by LC. Finally, this separation (efficiency $\approx$ 1500 plates) has not been, and almost certainly never will be, obtained by thin-layer procedures with corresponding speed and precision of quantitation. This work illustrates the close analogy in terms of performance, speed, and convenience of high-efficiency column LC and GC and its superiority to TLC.

If we turn to analysis of steroids, the nature of the challenge of LC to the established order of things, i.e., the use of GC, can be examined. Analysis of urinary steroids is becoming widely used, for example, in physiological studies of pregnancy. Some steroids are thermally unstable and the presence of strongly polar groups in their molecules make their gas chromatographic handling difficult, frequent recourse to derivatization being necessary.

Figure 8.2 shows an LC separation of six steroids (without derivatization) in about 15 min with an efficiency of about 120 plates per foot[3]. Despite the very small amount of work which has yet been done by high-efficiency LC in this field, it is clear it is already becoming competitive in terms of speed and performance, and is probably superior in terms of solute stability, to GC.

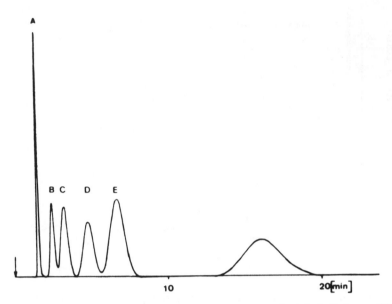

Fig. 8.2. Separation of steroids [reproduced by courtesy of J.A.R.J. Hulsman ([3])].

Another field in which LC looks likely to challenge GC is in the separation of pesticides, for much the same reasons as have already been mentioned–in particular, the reactivity of these materials in the very hot GC columns needed for their separation in acceptable times.

Figure 8.3 shows a separation([4]) of four common insecticides by LC on Corasil. The analysis time is about 45 sec and the efficiency around 100 plates per foot. In Fig. 8.4 some herbicides are seen separated by LLC

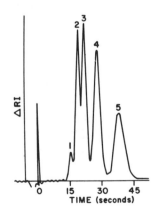

Fig. 8.3. Separation of insecticides by high-speed liquid chromatography. 1, aldrin impurity; 2, aldrin 6 μg/μl; 3, p, p′ DDT 6 μg/μm; 4, DDD 8 μg/μl; 5, lindane 10 μg/μl (Reproduced by courtesy of Waters Associates, Inc.).

RECORDER RESPONSE

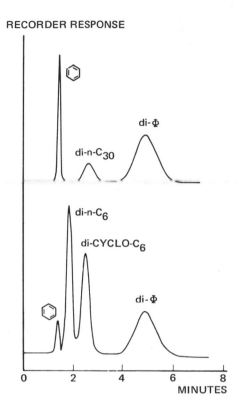

Fig. 8.4. Separation of phthalate esters by supercritical chromatography (reproduced by courtesy of *Analytica Chimica Acta*).

with oxydipropionitrile supported on the CSP support of Du Pont[5]. Here the analysis time is just a few minutes and the column efficiency nearer 200 plates per foot.

Supercritical chromatography has been pioneered by two Dutch workers, Sie and Rijnders[6, 7]. Almost all the published work is due to them, but, despite the experimental problems involved, there is evidence that the

Table 8.1. Critical Constants of Suitable Mobile Phases

Substance	T_c, °C	P_c, atm
Carbon dioxide	31.1	73
n-Pentane	196.6	33.3
i-Propanol	235.3	47
Diethyl ether	193.6	36.3

Table 8.2. Characteristics of Supercritical Separations

Enhanced volatility of solutes
Influence of mobile phase
Type, not size, separations
Large temperature effect
Low viscosity of mobile phase

potential they have shown the technique to possess is attracting other laboratories into the field.

The essential difference between gas and supercritical chromatography is that whereas in the fomer technique inert gases at relatively low pressures (up to a few atmospheres) are used, in the latter a variety of materials in the gas phase above their critical point can be used as mobile phase. Table 8.1 lists the critical points of some materials which have been used for this purpose.

It is immediately apparent that these substances cover a wide polarity range. In consequence, a very wide variation in mobile–stationary phase pair selectivity is available to superimpose on the volatility and solute–stationary phase interactions governing conventional gas chromatographic separations. In fact, supercritical chromatography has much in common with both liquid and gas chromatography. Let us make the appropriate comparisons.

The technique is still at a rudimentary stage and less well-developed than the high-efficiency LC previously discussed. In general there is a requirement not only for high-pressure but also for high-temperature operation which, on the one hand, makes for experimental problems and, on the other, for only limited applicability to labile materials. Supercritical techniques are far behind GC at present in instrumental sophistication and ease of use and it is probable that this will always be the most difficult area in which to work. Nevertheless, it may still prove to be superior to LC in some important respects. This is because supercritical fluids have much lower viscosities than liquids so that high flow rates through the very fine particles of column packing (needed to maximize efficiency) are much more easily attained.

Despite the practical difficulties, a considerable number of fascinating applications have been published; indeed in some respects it has established a lead over LC and, in some instances, over GC. It is well worth looking at the unique characteristics of supercritical separations. These are summarized in Table 8.2.

The first two factors are related; the interaction of the mobile phase with the solute leads to its much more rapid elution than with an inert gas at normal (i.e., gas chromatographic) pressures. For example, the retention times of benzene and pentyl benzene on a 1-m column of Porapak Q at

200° C are 7 and 150 min, respectively, with helium at 2 atm (i.e., normal GC practice) but 1 and 4 min with *n*-pentane at 26 atm. Remarkably, the same column retains dodecyl octadecyl benzene (C_{36}) for only 2 min 40 sec with *n*-pentane at 42 atm!

   Figure 8.4, taken from Sie and Rijnders' work([6]), dramatically illustrates the ability to achieve type, rather than size, separations. The column contained polyethylene glycol 6000 as stationary phase and *n*-pentane was the supercritical mobile fluid. In the upper chromatogram di-*n*-tricontyl phthalate is eluted well before the much smaller, more volatile diphenyl phthalate (the selectivity of PEG of 24 carbon atoms/benzene ring here in favor of the aromatic can be compared with a maximum achievable in GLC of about 8 carbon atoms). The lower chromatogram shows that the hexyl phthalate is eluted well before its cyclohexyl counterpart, which in turn has

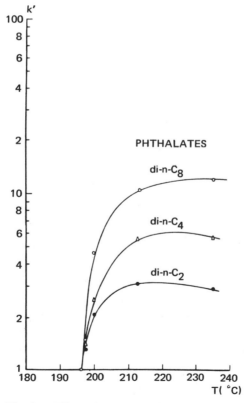

Fig. 8.5. Effect of temperature on retention in supercritical chromatography (Reproduced by courtesy of *Analytica Chimica Acta*).

a retention close to that of the C_{30} alkyl phthalate. Thus from these model compounds it can be seen that selectivities of 10 or more carbon atoms are achievable so that type groupings can be obtained on wide-boiling fractions, with boiling ranges of as much as 200 °C; This is well beyond the capabilities of GC, although similar to the effects in liquid chromatography.

Figure 8.5, also from Sie and Rijnders([6]), shows the very large effect of temperature on the relative retentions of di-n-alkyl phthalates just above the critical temperature of the eluent, n-pentane. It will be noted also that as the temperature increases from T_C to about 220° C the absolute retention also increases, the opposite effect to that which occurs in GC.

Thus supercritical chromatography is seen to achieve remarkable separations quite beyond the possibilities of GC. The real competition to supercritical chromatography comes from high-efficiency column LC and at the moment no truly objective comparison can be made. The latter is presently in the lead in terms of instrumentation and this is particularly so considering the choice of detectors which are available. Only UV and transport-ionization detectors have been successfully used with supercritical mobile phases and work has been restricted almost wholly to separation of aromatics. However, this lead is unlikely to be long-lived as more effort is devoted to the newer technique. An advantage LC has which will not be challenged is its applicability to very high-molecular-weight substances, including high polymers. Thus LC appears likely to have the edge in terms of scope, relative mildness of operating conditions (especially temperature) and simplicity and safety of instrumentation, but the very fast analyses demonstrated with the other techniques make it an attractive proposition for those materials which are just beyond the reach of GC.

The final comparison to be drawn is between column and thin-layer liquid chromatography. To some extent this has already been covered earlier in the book by comparing the rate of generation of theoretical plates, showing at least a 1–2 order of magnitude advantage in favor of the column. Thin layers are cheap, easy to use, quite fast, and fairly efficient, whereas columns can be costly and they are sophisticated, very fast, and very efficient. Columns should be used for analytical purposes when high resolution ($> 5 \times 10^2$ plates), fast analysis (< 10 min), or precise quantitation is required.

In conclusion, gas and thin-layer chromatography will remain the most generally useful analytical control techniques, but column LC will become increasingly important because of its potential for accurate, automatic analysis with high resolution.

REFERENCES

1. Porter, R. (ed.), *Gas Chromatography in Biology and Medicine,* J. and A. Churchill, London, p. 154, (1969).

2. Horvath, C. G., Preiss, B. A., and Lipsky, S. R., *Anal. Chem.* **39,** 1422 (1967).
3. Hulsman, J. A. R. J., Thesis, Univ. Amsterdam (1969).
4. *Chromatography Notes* (Waters Associates, Inc. Framingham,, Mass.) **1** (1), 4 (1970).
5. Kirkland, J. J., *J. Chromatog. Sci.* **7,** 7 (1969).
6. Sie, S. T., and Rijnders, G. W. A., *Anal. Chim. Acta* **38,** 31 (1967).
7. Sie, S. T., and Rijnders, G. W. A., *Gas Chromatography 1968*, Harbourn, C. L. A., ed., Inst. Pet., London, p. 235, (1969).

Author Index

Numbers in parentheses refer to references at the end of the chapters and to further reading.

Subject Index